MOONSTRUCK

Moonstruck

HOW LUNAR CYCLES AFFECT LIFE

ERNEST NAYLOR

OXFORD
UNIVERSITY PRESS

OXFORD
UNIVERSITY PRESS

Great Clarendon Street, Oxford, OX2 6DP,
United Kingdom

Oxford University Press is a department of the University of Oxford.
It furthers the University's objective of excellence in research, scholarship,
and education by publishing worldwide. Oxford is a registered trade mark of
Oxford University Press in the UK and in certain other countries

First Edition published in 2015

Impression: 1

Published in the United States of America by Oxford University Press
198 Madison Avenue, New York, NY 10016, United States of America

British Library Cataloguing in Publication Data
Data available

Library of Congress Control Number: 2015934015

ISBN 978–0–19–872421–6

Printed in Great Britain by
Clays Ltd, St Ives plc

For Elizabeth and Helen

CONTENTS

PREFACE

'Tell me Silent Moon, what are you doing in the sky, Silent Moon?'   Giacomo Leopardi (1798–1837).

'Beliefs . . . in an influence of the Moon on life on Earth . . . are by no means all moonshine.'   H. Munro Fox, 1928.

As an introduction to possible effects of the Moon on living organisms, nothing could have been more effective for me than a memorable event that occurred during my undergraduate days. At that time the Student Biological Society of the University of Sheffield was fortunate enough to lure Professor (later Sir) Alister Hardy, a distinguished marine biologist from Oxford University, to deliver an invited lecture. Fully expecting a lecture on his research concerning ocean plankton, the Committee delegated to me, as an aspiring marine biologist, the task of delivering the vote of thanks to the speaker after his talk. Duly flattered, but apprehensive, a short refresher course on Hardy's pioneering studies of instrumentation for collecting ocean plankton continuously from moving ships, and experiments on the nightly upwards migrations of plankton, seemed necessary for appropriate preparation. However, all was in vain when the eminent Professor began his lecture by announcing that he would not speak about marine plankton, but about what he called 'aerial plankton', based on research he had carried out a few years earlier when he was Professor of Zoology at the then University College of Hull. He delivered a fascinating lecture, discussing day and night collections of insects, captured in various weather conditions and states of the Moon, using tow nets that he had arranged to have attached to railway trains travelling between Hull and London. As a lesson in how to think on one's feet or, more appropriately, one's seat, it was a salutary experience as my pre-prepared vote of thanks began to crumble away. Needless to say, I heard some of the lecture and certainly began to appreciate that living organisms may be affected by the light of the Moon and not simply by light changes between day and night. Others, before and after that time, have considered whether or not aspects of the behaviour of animals are affected by the Moon and the question to be addressed in this book is the extent to which there is scientific evidence that the Moon has influenced living organisms, particularly animals, including humans, throughout the timescale of evolution.

* * *

We live in an age when the reality of the Moon has been studied intensively by remote-sensing from orbiting, man-made satellites, and by instruments that have been engineered to soft-land on the Moon after rocket-propelled transport from Earth. Impressively, too, sophisticated gadgetry has been placed there by astronauts on Apollo missions. Contrast these events with perceptions of the Moon by early humans, for whom cyclical changes in the size, shape, and position of the Moon in the sky had mystical properties. Such properties were formalized in lunar myths and legends among citizens of societies worldwide, as in those of ancient Rome and Greece, who considered, for example, that sheep's wool and human hair grew more rapidly when the Moon was in the ascendant. From earliest times, such beliefs have persisted, even concerning god-like influences of the Moon on various aspects of life on Earth, beliefs that in some cases persist until the present day, not least in astrology. Yet juxtaposed alongside such beliefs have been the views of realistic sceptics, who deny the possibility of any linkages between life processes on Earth and changes in the appearance and position of the distant Moon. Indeed, matters relating to the Moon are so widely viewed with scepticism that some even question the veracity of the extensively televised lunar landings by American astronauts, proposing that they were no more than elaborate hoaxes.

The word 'lunacy' is embedded in the English language. On the one hand it implies linkages between some aspects of human behaviour and phases of the Moon but, on the other hand, it can be used to describe the 'lunatic' state of mind of individuals who postulate that such linkages do occur. Over many years, until the present day, a plethora of contrasting 'beliefs' and 'disbeliefs' have extended to so-called Moon-related phenomena in many aspects of life on Earth, no doubt partly fostered by the imagination and creativity of artists and writers throughout history. Accordingly, scepticism concerning Moon-related phenomena, arising from a long legacy of myths, legends, and artistic imagery, is deeply entrenched in the human psyche. Indeed, even the scientific study of Moon-related biological phenomena has been slow to develop, with scientists fearing ridicule at the prospect of entering the field. This is in spite of the fact that in recent decades biological adaptation to solar cycles is seen to be commonplace. It is widely recognized that animals and plants are able to continue with rhythmic behaviour approximately in phase with day and night even when they are isolated in constant conditions. It is understood that such behaviour is controlled by internal biological clocks of circadian period-icity, understanding that is reinforced by human experience of jet-lag during travel across time zones. There is good understanding, too, that humans possess internal circadian clocks, the timing of which can be disrupted, to the detriment of well-being and even health.

It has taken fifty years or so for the study of heritable circadian clockwork within the science of chronobiology to become fully accepted, but the molecular basis of the clock-work, in the form of identifiable clock genes, is now generally understood. It is also generally recognized that the clock genes are involved in the release of substances, such as melatonin and cortisol, which mediate in the control of biological rhythms at the level of the whole organism. In humans, reliant as we are on manufactured clocks and watches, we are not always aware of our biological internal timing system, but we certainly become aware of it when it is disturbed by jet travel. At such times we also recognize that not all our physiological systems run on the same time. We may adjust our sleep patterns after some days in a new locality after extensive journeys east or west, but our digestive and cognitive systems may take a little longer to catch up.

Chronobiology is also sufficiently advanced now that the medical profession is coming to recognize the significance of our inner biological timing system. There is now acceptance that circadian clock disruption can lead to metabolic and behavioural problems such as obesity and depression, as well as to more commonly understood symptoms of jet-lag. There is also increasing acknowledgement that the efficacy of prescription medication may vary according to the time of ingestion or application. Accordingly, a related science of chronotherapeutics is emerging whereby the optimal times for drug application are being determined from basic knowledge of the human circadian clock system.

In contrast to our understanding of circadian biological rhythmicity, matching the solar day, our understanding of possible Moon-related—that is, circatidal and circalunar— biological rhythmicity has lagged seriously behind. Nowadays, however, cushioned by the fact that it has been scientifically established that some living organisms have adapted to *indirect* effects of the Moon through its influence on tides, even *direct* biological effects in response to, or in anticipation of, changing moonlight are being characterized and gaining in acceptability. In this book I will discuss these newly emerging aspects of the science of chronobiology, with examples from plants, but particularly among animals, with reference also to supposed lunar influences on human behaviour, setting them against the mythologies attributed to the Moon throughout human history.

Repeatedly throughout history supposed Moon-related effects upon human life processes have been critically dismissed with understandable scepticism. Recently, however, evidence is building up, for a small number of animals, that clock genes related to the periodicities of the Moon and tides may be present in their genetic make-up alongside the more familiar circadian clock genes. If circatidal and circalunar genes are present within some living organisms, might they be more widespread in the genetic make-up of animals and plants, and even in humans? If they are, perhaps the search for Moon-related rhythms in humans should now gain more respectability. At least one very recent study indicates that humans may indeed be inherently responsive to moonlight in the determination of patterns of sleep, paving the way for further scientific enquiry into the relationship between the Moon and the human condition. Humans have obviously adapted to the patterns of the solar day; are we now seeing that, less obviously, they have also adapted to the lunar day and month? If so should we at least remain open-minded that moonlight may have more of an impact on our lives than few in modern societies have previously been prepared to admit?

* * *

Several people have kindly read and commented on all or parts of drafts of this book, or have otherwise contributed with help and discussion at various times, namely: David Bowers, Elizabeth Galloway, Michael Hastings, Helen Hawkes, Elfed Morgan, Gillian Naylor, Graham Walker, and Simon Webster. To all of these I am extremely grateful for their time, patience, and encouragement, as I am also to two anonymous referees who appraised my original proposal so constructively and enthusiastically for OUP. At Oxford University Press I am most grateful to Emma Ma, Jenny Nugee, Kate Gilks, and colleagues for their considerable help and advice on the preparation of the book. Above all I thank my editor, Latha Menon. Her guidance in writing for a general readership, and her helpful questions concerning some of the science, have made for very constructive inputs to the text. Finally I thank my wife Gillian for her continuous support and critical reading of drafts, together with her patience and understanding when the concept of retirement took on a new meaning. Needless to say, I accept responsibility for any errors that may be apparent in the book.

Small superscript numbers in the text refer to the numbered list of References that follows the Glossary.

LIST OF FIGURES

1

MOON MYTHS
AND LEGENDS

Curiosity about the changing faces of the Moon is deep in the human psyche and at various times in human history there have been many supposed links between lunar periodicity and the natural world.

Not least, there are long-established perceptions that aspects of human behaviour are influenced by the Moon, which, worldwide, has in the past been worshipped as a deity. In Roman times the Moon was deified through the Moon-goddesses Juno and Luna, based on earlier Greek manifestations, and was endowed with powers that influenced philosophical and scientific thinking. The Moon's influence was characterized in the writings of the Roman Stoic philosopher Seneca, who argued that the Moon existed to make humans feel good by measuring time and making corn grow in an orderly way.[1] He was one of many scientific philosophers of the time who had a good knowledge of the Moon and its movements around the Earth, and was able to relate those movements to the occurrence of tides and eclipses. Perceived associations between the Moon and human behaviour are evident, too, in the writings of Tacitus, a century later, who stated that the Germans held their 'political' meetings during times of the full Moon or new Moon, regarded as times most favourable to start new initiatives. But the perceived association between the Moon and human lifestyle is much more pervasive than this, probably going back to the dawn of human thought.

It is believed that, even in Palaeolithic times, estimation of the passage of time during pregnancy was determined by observation of the lunar cycle, which approximated to the menstrual cycle. Some evidence for this comes from a French cave wall relief dated around 22,000 BC that has been interpreted as including a pregnancy calendar.[2] The image depicts a pregnant woman holding a crescent-shaped animal horn with thirteen notches, the approximate number of lunar months in the solar year (Fig. 1).

This is a highly subjective interpretation of an ancient artefact, but early rationalization of a year of thirteen lunar months survived in European peasant communities well beyond the end of the first millennium AD. It has even been suggested that lingering superstition of the number thirteen in modern times relates to inadvertent recollection that the thirteenth lunar month is the last, or death, month of the year.[3]

The repeated waxing and waning of the Moon no doubt gave rise to many beliefs among ancient civilizations, not least

Fig. 1. A cave wall relief, the
Venus of Laussel, from the
Dordogne region of France,
dated around 22,000 BC.
The pregnant woman holds
an animal horn with 13
notches, interpreted today
as a pregnancy calendar.

in relation to the Moon's influence on menstruation, birth,
growth, ageing, and death. For many, the repetitive nature of
the lunar cycle was perceived to be evidence of the divine
nature of the Moon itself. As Robert Graves wrote in *The
Greek Myths*:

> Time was first reckoned by lunations, and every important cere-
> mony took place at a certain phase of the Moon. (i. 13)
>
> ... the religious system of the Neolithic and Bronze Ages in
> Europe seems to have been remarkably homogeneous, being
> based on the same mystical relationship between the Moon-
> goddess and her sons (i. 11)

3

Indeed, perceptions of lunar deities were so entrenched in human thinking that, by the time of the Mesopotamian civilization of the third century BC, the Moon was considered to be the dominant sky god particularly concerned with the rhythms of life.[4] Relatedly, the Babylonian goddess Annit, superseded by Ishtar, was originally believed to rule the Moon. Subsequently, the imagined influence of a lunar deity was so widespread that it was conceptualized in pre-Greek and Asian societies as the Moon-goddess, Selene, whose influence persisted into the time of classical Greek culture. The Selene mantle was later assumed by Artemis, regarded by some as a nature goddess, particularly of forests and hills. Throughout the Greek world different cults endowed particular gods with different talents but, to many, Artemis was a major Olympian goddess who presided over crucial aspects of human life processes and rites of passage, including childbirth. The medicinal herb *Artemisia* (mugwort) was so named because it was used to try to induce delivery during the birth of a child.

Lunar influence on early societies was such that, from earliest human history, the passage of time was not only estimated by monthly cycles of the Moon, but annual solstices and equinoxes, which could not be determined exactly, were approximated to the nearest full or new Moon. Originally the year was divided into Moon cycles, hence 'months', rather than fractions of the solar year. However, even when it was later established, after astronomical observations, that the solar year consists of 364 days, with a few hours to spare, it was divided into so-called Moon cycles. The number 364 conveniently divided into thirteen 'months', each of 28 days, numbers which became sacred. Significantly, too, the timing of the 28-day 'month' approximates to the timing of the menstrual cycle of women, enhancing the case for the Moon to be worshipped as a female goddess. This supposed linkage was challenged long ago by Aristotle, who attributed it to coincidence, but the approximate conformity was given some credence much later by none other than Charles Darwin. In his writings Darwin asked whether, since humans are descended from fish-like ancestors, the approximately 28-day feminine cycle might not be a vestige of the past when life depended on the tides and therefore on the Moon.[5] In fact the so-called 28-day 'month' is an artificial construct. As we shall see in Chapter 2, the interval between successive full Moons, seen at the same position in the sky, from the same place, is 29.5 days. This is the duration of the synodic or true lunar month, the observed pattern of lunations which ancient (and modern) observers on Earth were characteristically aware of. It differs from estimates that the mean period of the human menstrual cycle is 27.32 days, to the extent that though a single episode of a typical menstruation cycle might coincide with the day of a full Moon, subsequent episodes would regularly change phase in relation to the full Moon, until they coincided again some thirteen or fourteen months later.

The occasional coincidences of particular women's menstrual cycles with specific phases of the Moon may in the past have fostered the myth of the Moon as a fertility goddess, but the timings occur by chance. We will consider in Chapter 9 whether there is any evidence that the approximate periodicity of the human menstrual cycle is an evolutionary vestige, as Darwin suggested, but for now we can provisionally conclude that Aristotle's view is correct.

Nevertheless, imagery of the Moon-goddess became even more sophisticated in Greek culture. The three phases of the Moon from new to full to old were envisaged as relating to three ages of women, from maiden to nymph (nubile woman) to crone, each attributed to separate Moon-goddesses, respectively Selene, Aphrodite, and Hecate, sometimes unified as Hera rather than Artemis. Different Greek societies had various names for their Moon-goddesses, and the concept of the triple Moon-goddess was common. Elsewhere in classical Greek mythology the Moon was consistently represented as Cynthia, again a triple deity.[6] Moreover the symbolism of the Moon appeared in many artefacts. In Cretan mythology a symbol of royalty was a double-bladed axe which has been interpreted to represent waxing and waning of the Moon. The hooves of horses, notably those of the winged horse Pegasus, also represented the Moon. Artemis, often depicted on a silver throne in the shape of a crescent Moon, was sometimes referred to as the Maiden of the Silver Bow, the bow again representing the new Moon.

* * *

Some Greek deities were later adopted under other names in classical Roman society, but notions of lunar deities pre-date the classical world and spread world wide. Early diasporas of human societies through China, Japan, and parts of the Pacific Rim, including Hawaii, Indonesia, and Polynesia, all led to dissemination of the inherent concept of Moon-goddesses particularly associated with fertility. The concept also spread with the colonization of North America across the Bering Strait from Asia more than 20,000 years ago, and then to South America, reaching as far south as Chile. In southern South America the Moon-goddess Auchimalgen served as protector of the Auracanian Indians long before European colonists and settlers arrived. Particularly powerful obeisance to lunar deities was also to be found in Mexico and central America, well documented for the Mayan civilization during the Classic Period in Mesoamerica (AD 250–900). The Maya had detailed mathematical knowledge of the appearance and movements of the Moon, which was considered an important goddess.[7] They were able to make accurate predictions of celestial events, linking them with the role of their lunar and other celestial deities (Fig. 2).

Their knowledge was based on interpretation of lunar and other celestial movements from observatories still to be found in archaeological sites today.

Mayan civilization originated in the first millennium BC in the forests of southern Mexico and spread northwards into Yucatan, where the Maya built great cities, including that of Chichen Itza, in the present-day ruins of which a purpose-built astronomical observatory can still be seen. At the height of

Fig. 2. Mayan image of the monster Tzitzimime, from the Dresden Codex, with representations of eclipses of the Sun (top left) and Moon (top right).

their power the Maya produced codices in the form of folded, illustrated books containing detailed calendars and almanacs. These contained imagery of lunar gods and amazingly accurate astronomical tables, including patterns of lunar cycles and the timing of eclipses. The most famous surviving document, the Dresden Codex, was probably acquired by the Spanish conquistadors and sent by Cortez to the Spanish king before eventually finding its way to the royal library in Dresden. The surviving document is thought to have been written in the eleventh or twelfth century AD as a copy of an original produced three or four centuries earlier. Even so, it is considered to be the oldest known document written in the Americas.

Towards the end of the tenth century AD the splendour and sophistication of the classical Mayan peoples began to collapse, presaged by the destruction by fire of the city of Teotihuacan. That city, a tourist site today, with its colossal pyramid temple structures to the Sun and Moon, and the Great Square of the Moon, was almost certainly an important religious centre for centuries. However, with the decline of Teotihuacan, the detailed astronomical and mathematical knowledge based on Mayan culture also began to decline, as first Toltec and later, in the fourteenth and sixteenth centuries AD, Aztec cultures predominated. Even so, the Toltecs enhanced some of the observatories built by the Mayans, including those at Chichen Itza and Uxmal, which permitted later Aztec astronomers to calculate the length of the solar year. Moreover, obeisance to Moon-gods persisted throughout that time; the Aztec

Moon-goddess, pictorialized with Moon-shaped nose orna-
ments, was believed to control many life processes, including
human sexuality, fertility, procreation, and growth, together
with rainfall and the ripening of crops. Such beliefs persisted
until the Spanish conquests, beginning with the arrival of Cortez
during the sixteenth century, and the spread of Christianity.

Unusually, one of the images in the codices is of a
Moon-goddess holding a rabbit and framed by the crescent
Moon. The rabbit on the Moon is an old belief in Mesoamerica,
perhaps relating to the growth of vegetation and the success of
crops such as maize. A story goes that in ancient times the Sun
and the Moon were of equal brightness so that it was difficult
to sleep at night. A Moon-god threw a rabbit towards the
Moon, wounding it and darkening it with the grey patches
that we see today. The Aztecs did not recognize the 'face' of the
Moon as that of a man, but of a rabbit. Notwithstanding the
long history of Moon myths and legends, there has been an
almost equally long history of rational consideration of such
beliefs which has rejected them as merely superstition. Even
Seneca, writing in Rome two thousand years ago, warned
against superstition when legend had it that eclipses of the
Moon were portents of disaster and were times when witches
were able to travel to the Moon.[1]

Seneca considered that superstitions, not eclipses, were the
true disasters. He wrote that his descendants would in due course
'be amazed' that his contemporaries did not know things that
should have been plain to them. Such rational considerations

of lunar mythology have inevitably built up over the years, to the extent that modern scepticism is rife that there are any associations at all between lunar phase and life on Earth.

Nevertheless, even in more recent history, the imagery of links between the Moon and human life processes has been perpetuated in art and literature.

* * *

Though the imagined Moon of ancient mythology was a female goddess, a traditional and geographically widespread personification dating from historical times is the 'Man in the Moon'. This interpretation of the observed aspect of the full Moon, with its not entirely convincing perceived shape of a human face, persists in stories told to unbelieving children to the present day. Based initially in folklore, the image is probably reinforced by a biblical story in which a woodcutter who refused to rest on the Sabbath was stoned to death and banished to the Moon. In another biblical version of the story the man punished by banishment to the Moon was Judas, who denounced Jesus Christ. The same image of the lunar face is common in religious art, not least in fourteenth-century frescoes in Padua, including those of Giotto in the Scrovegni Chapel.[8] Yet earlier mythical stories named the Man in the Moon as Endymion, who was taken to the Moon by his lover Selene and who was named by Pliny the Elder as the first individual to observe the motion of the Moon.

This dichotomy of male and female imagery of the lunar surface persisted until at least the sixteenth century. In England, for example, the cult of the Greek Moon-goddess was intensified in the context of Queen Elizabeth I, the Virgin Queen, when in the 1580s Walter Raleigh published a poem entitled 'The Ocean to Cynthia', reflecting his Queen's unapproachability. Moreover, as William C. Carroll wrote in 2001, 'Elizabeth actively promoted herself as the chaste, distant, lunar Cynthia', suggesting too that Shakespeare mocked the Queen's perceived image in his comic play *A Midsummer Night's Dream*, published around 1595.[6] The play takes place largely under moonlight during the shortest night of the year at the summer solstice, juxtaposing its main imagery of the Moon as Cynthia with allusions to Elizabeth, and referring to the Man in the Moon in the play within the play.

Indeed the phrase 'Man in the Moon' became so entrenched in European thinking that the *Oxford English Dictionary* came to refer to it as an old reference for a 'pretendedly unknown person who supplies money for illicit expenditure at elections'.

More recently the idea of the Man in the Moon was exploited for amusement in George Melies' 1902 science fiction film *A Trip to the Moon* in which a 'rocket' carrying a group of astronomers fired at the Moon becomes partly impaled in the eye of the Man, from which molten cheese is seen to drip as a simulated tear (Fig. 3).[9]

Melies' astronomers, still carrying umbrellas, then proceed to meet the locals, called Selenites. Encountering problems with the residents, the astronomers eventually return to the safety of their spacecraft, which they somehow manage to push off the edge of a cliff, and fall safely back to Earth.

Fig. 3. An early space film Moon-landing by a 'rocket'-borne space capsule fired from a cannon.

Thereafter, until the 1960s several Moon-based movies continued to foster lunar myths. They persisted with historical misapprehensions, some no doubt deliberate for artistic effect, including depictions of jagged mountains on the observed face of the Moon, assumptions of the occurrence of a lunar atmosphere similar to that on Earth, and indications that lunar gravity was also equivalent to that of the home planet. Some films also fostered Melies' myth that an object could fall off the surface of the Moon and be attracted back to Earth by the pull of terrestrial gravity. The presence on the Moon of weird plants and animals and, in some cases, even scantily clad women were further artistic liberties taken by early film-makers.

In early historical times the observed face of the Moon was probably considered to be relatively flat and featureless except for the dark patches that gave rise to mythological interpretations of 'The Man in the Moon' or 'The Rabbit on the Moon'. With the advent of the telescope and photography the old perceptions altered little, except that the presence of lunar craters came to be recognized. Nevertheless, myths of jagged mountains, as presented by movie makers, persisted even in the early 1950s. It was Galileo who, in the late autumn of 1609, first reliably sketched the face of the Moon as viewed through his newly constructed telescope, an instrument invented and patented by Hans Lippershey in The Hague a year earlier. That is not to say, however, that naturalistic images of the Moon in art were unavailable before that time. As early as the fourteenth and fifteenth centuries AD artists were beginning to paint the Moon naturalistically, instead of in the stylized imagery of earlier times. It has been suggested that many artists of those times, such as Giotto and Jan van Eyck, themselves made revolutionary advances in pre-telescope astronomy to produce more accurate representations of the Moon as part of their observations and depictions of the natural world.[10,11]

The Moon has mesmerized others too, including the mathematician Thomas Harriot, who was the first to view the Moon through a telescope. He did so a few weeks ahead of Galileo, but the observations and drawings of Galileo were the starting point of lunar mapping, developed later by Gian Domenico Cassini (1625–1712), Tobias Mayer (1723–62), and Joseph Johann von Littrow (1781–1840).[12] Their maps are classics, all improved by observation of the Moon during eclipses, when lateral shading of sunlight enhances the shape of lunar craters. Also, not surprisingly, particularly naturalistic images of the Moon began to appear in paintings soon after the discovery of the telescope. Notable examples are in the works of Donato Creti, who, in 1711–12 produced a series of paintings entitled *The Astronomical Observations: The Moon*. In one of these an observer is seen viewing the Moon through a newly developed telescope, the lunar surface having been painted by Raimondo Manzini in an obviously relatively enlarged and inverted image, exactly as would be seen through an early telescope.[11] This and other such paintings were commissioned by Count Luigi Ferdinando Marsili, an amateur astronomer, in what turned out to be a successful attempt to persuade Pope Clement XI to support the construction of an observatory in Bologna to house Marsili's astronomical observing instruments. Other important artists who produced notable realistic images of the Moon included John Russell, RA (1745–1806) and James Nasmyth (1808–90), son of the Scottish landscape painter Alexander Nasmyth, all again with the aid of the telescope.

If the telescope changed scientific attitudes to the nature of the Moon, it also changed literary styles.[6] Prior to the early seventeenth century, literature concerning the Moon was heavily centred on the personification of mythological stories, as in

Shakespeare's cloaking of Queen Elizabeth I with the mantle of the goddess Cynthia. Afterwards, as in films later, with detailed observations of the Moon, new forms of imagination began to hold sway. The discoveries of Galileo stimulated new forms of literature, since the newly observed cratered lunar surface could be interpreted to indicate a terrain of mountains, seas, and rivers. The nomenclature of perceived 'seas' (Maria) and 'ocean' (Oceanus Procellarum) persists today, describing dark patches on the lunar surface that are plains of basalt rock formed by ancient volcanic eruptions. At the time, too, notions of voyages to the Moon began to be developed, and possibilities for human colonization were imagined. Indeed, following Galileo's descriptions of the face of the Moon in the early seventeenth century, Francis Goodwin, then Bishop of Hereford, wrote a derivative book entitled *The Man in the Moon*. In it he imagines an intrepid explorer who harnesses a flock of geese to fly to the Moon. Landing, after more than a week of flight, the explorer discovers a mighty sea, with projecting islands, in the shape of a dark patch seen from Earth. Lunar features are described as enormous when compared with similar objects on Earth, with trees three times their size back home, inhabitants twice the size of earthlings, and even doorways of buildings described as rising to thirty feet in height. Around the same time John Wilkins, a mathematician who much later became one of the founders of the Royal Society of London, summarized available scientific knowledge concerning the Moon in *The Discovery of a World in the Moon*. In his interpretation the dark patches on the Moon also represented seas and brighter parts the land. He even estimated the distance between the Earth and the Moon as 179,172 miles, a figure so precisely and authoritatively stated that it persisted in other publications for many years until modern estimates became available. Nowadays the estimate is rather larger, with an average distance of 238,857 miles, which is greater or less according to the elliptical nature of the Moon's orbit. Wilkins, taking a strictly scientific approach, also considered the nature of what would be required to escape the Earth's gravity and the temperature changes that a lunar voyager might encounter. Therefore, from the time of Galileo's telescopic observations of the Moon, early myths and legends concerning perceived lunar influences on the Earth were challenged in favour of the concept of the Moon as a separate new world, despite mockery from some more conservative literature at the time. Then in 1687 Isaac Newton published his *Principia Mathematica* explaining the gravitational interaction between the Moon and the Earth and a scientifically determined influence of the Moon on the Earth was established, most obviously seen in the rise and fall of ocean tides.

Visits to the Moon to dispel earlier myths were no doubt unimaginable at the time, except in fiction. Nevertheless, as Gian Domenico Cassini, intent on pursuing scientific study of the Moon, wrote to colleagues at the Académie Royale in Paris, in 1692:

That is the advantage of lunar maps. Those who do not look at things in depth, think that they are useless descriptions of an imaginary country. They are surprised that people of common sense enjoy themselves making such exact maps of a lunar world where certainly nobody will ever go, whether to make conquests or to found colonies.[12]

For over a hundred years after Cassini made his case for studying the Moon 'because it is there', astronomers continued to record their observations by drawing, but, by 1840, the first photographs of the lunar surface were made, a technique that is still in use today. The precision of lunar photographs in the 1850s was already very high and full-frame images of the Moon have scarcely improved since. Preceded by photography, the drive to elucidate the true nature of the Moon's surface then began, followed by Moon-targeted rocketry and satellites, culminating in the manned lunar landings.

Yet, despite all the photographic imagery available, film-makers up to the 1950s continued to foster older myths that the observed face of the Moon was jaggedly mountainous. It was not until Stanley Kubrick's 2001: A Space Odyssey[9,13] in 1968 that the relative smoothness of the visible cratered surface of the Moon was portrayed in film. It seems that Kubrick had the benefit of the first ever in situ photographs of the lunar surface that were beamed back to Earth from a Soviet Robotic Lander despatched to the Moon by rocket from the USSR. The unmanned Soviet spacecraft Luna 9 landed safely on the Moon in the region of Oceanus Procellarum in February 1966, but even then the world may have inadvertently been misinformed about the lunar surface.[14] Amusingly, the Soviets did not release their images immediately, but signals from their lunar Lander were intercepted by the British Jodrell Bank radio telescope. There, apparently, the data were recorded on a standard fax machine, such as those used at the time by newspapers, and the pictures were released before the official release by the Soviets. Unfortunately the wire-service fax machines compressed the photos laterally by a factor of 2.5, thus exaggerating even more the low sun angle of the pictures to suggest that the observed lunar landscape was indeed mountainous and jaggedly so. Then, to exacerbate the unfortunate occurrence, a geologist commenting on the 'scoop' photographs suggested that the 'mountains' were indicative of volcanic activity on the Moon that might indicate the presence of precious metals. This statement then led US newspapers to claim that Luna 9 had discovered a vein of gold on the Moon. The Soviets enjoyed the Western gaffe before publishing the real photographs of a lunar surface with the gentle hills and rolling slopes associated with lunar craters.

In the twenty-first century it turns out that the far side of the Moon, as distinct from the normally observed near side, does indeed have a landscape of jagged mountains, the appearance of which the USA continues to survey with specially orbiting satellites. So, with the pioneering unmanned space programmes of the Soviets and the USA, the spectacular manned touchdowns of the US Apollo programme, and increasing numbers of lunar satellites and landers launched by other countries around the world, the imagined surface of the

Moon has become reality, dispelling myths and raising possibilities of future exploration and exploitation by man. Already a Lunar Exploration Society has been founded by a serious grouping of scientists, technologists, resource managers, and environmentalists, partly under the auspices of the European Space Agency. The Society was established with the declaration that 'Development of human capability on the Moon will be the next step in humanity's emergence into the Universe'. This development is a far cry from the perceptions of French Academicians in the late seventeenth century who challenged the need to make maps of the surface of the Moon where 'nobody would ever go'. In a salutary event for sceptics of 'blue skies' research Neil Armstrong gave the lie to that prediction when he became the first human being in history to set foot on the Moon on 21 July 1969. Moreover, even before he touched down, as he piloted the lunar module to a suitable landing place on the Moon's surface, stirring up clouds of lunar dust as he did so, he was also dispelling the childhood myth that the Moon was made of green cheese.

Not all ancient myths, however, have been dispelled in modern times. From Roman times it has often been considered that the growth of plants is affected by the phases of the Moon. So much so that since the early twentieth century methods of so-called 'biodynamic' agriculture and forestry have been developed worldwide, which, even in modern times, purport to take account of supposed relationships between optimal plant growth and lunar cycles.[15,16] In forestry, terms such as

'Moonwood' have even been created which imply that there are favourable times of the lunar cycle when wood should be cut, depending on the use to which the timber is to be put. Some of these notions are perpetuated in folklore relating to the choice of timber for the construction of musical instruments. Many also have echoes of a Royal Forest Order issued during the reign of Louis XIV (1643–1715), which stated that felling of trees should take place during a waning Moon in winter, presumably related to the supposed optimization of timber to be used for building ships for the French navy. Around 1734 the order was challenged by M. Duhamel du Monceau, a pioneering tree biologist who was General Inspector of the French Navy at the time and who believed that the order was based on superstition. He carried out so-called 'experiments', the outcome of which he claimed contradicted the rationale of the order, but then went so far as to claim that the optimal time for felling timber was during the waxing phase of the Moon! So far there is no convincing evidence that the claims of 'biodynamics' concerning forestry or agriculture are scientifically justified—claims that are not helped when some of the recommended planting times for optimal growth have connotations of astrology. However, as we shall see in later chapters, there is good evidence that the Moon affects the behaviour of some members of the living world, notably many animals. First, though, it is necessary to consider the nature of the physical environment in which animal adaptations to lunar cycles seem to have taken place.

2

THE BIG SPLASH

In November 2000 an international conference was held to discuss the current state of knowledge of Earth–Moon relationships. The conference was held to celebrate the 400th anniversary of the founding of the Galilean Academy of Sciences, Literature, and Arts at the University of Padua, in Italy, where Galileo made his pioneering breakthrough in the field of astronomy at the start of the seventeenth century. The conference was opened by a contemporary Professor of Astronomy in Padua, Cesare Barberi, in the superb Aula Magna of the University Palace in Padua, where Galileo taught. In his opening address Professor Barbieri noted that

The Earth harbours life, the Moon is extremely sterile, however our natural satellite regulates the life, through the stability of the terrestrial rotational axis, through the tides, and perhaps through subtler effects that still need to be better understood. Therefore the study of the Moon and of its relationships with the earth should be of highest priority.

The conference heeded that plea by bringing together academics of many disciplines including cosmology, astronomy, physics, chemistry, geology, climatology, biological sciences, and the arts, in company with an astronaut, Commander David Scott, who, in the Apollo 15 mission, had walked on the Moon. Together they honoured the memory of Galileo, appraising the myths and realities of Earth–Moon relationships as they were understood four hundred years after Galileo excitedly first sketched what he saw of the lunar surface through the newly invented telescope.

* * *

The story of the relationship between the Moon and the Earth can be traced back to a time billions of years after the Big Bang, which occurred about 9.5 billion years ago and from which matter formed.[17] This matter eventually became stars in great galaxies within which were nebulae—massive gas clouds filled with debris from dying stars. Within one such nebula, probably triggered by energy from a dying star in the form of a supernova, a huge cloud of gas and debris was destined to coalesce to become our solar system. Within the swirling debris around the nascent Sun, the Earth took shape some 4.6 billion years ago, though at that stage it lacked its own orbiting satellite—the Moon. How the Moon was formed has for a long time been a matter of speculation.

Reflecting the deified influence of the Moon on early Greek societies, it is not surprising that Greek science at the time sought explanations for its origin. It was probably Thales, in the sixth century BC, who first maintained that the Moon had no light of its own but receives and reflects it from the Sun, on the grounds that it carries little of the warmth and brilliance of sunlight. Yet Hellenistic Stoic philosophers from about 300 BC considered the Moon to be a globe of fire and air.[18] Later classical Greek science had a less nebulous explanation for the nature of the Moon, referring to it as a 'terrestrial body', and repeated references in the writings of Greek philosophers, including Plutarch (AD 50–120), describe the Moon as having 'an earthly nature', a notion that persists in scientific explanations of the Moon's origin today. In his treatise entitled *On the Face which Appears on the Moon*, from around AD 90, a rare surviving account from classical antiquity especially concerned with lunar theory, Plutarch also recognized that moonlight is reflected sunlight. He argued against Aristotle's view that the Moon, like other planets and stars, was a perfect and divine being. Later Greek philosophers challenged popular beliefs that lunar and solar eclipses were terrifying events that presaged misfortune and disaster, offering countervailing rational arguments against deep-seated lunar myths and legends. Nevertheless, despite the groundswell of rationalism by those Greek philosophers, Aristotle's view of the Moon and other planets and stars as divine beings, that orbited around the Earth, prevailed substantially until the time of Galileo at the beginning of the seventeenth century. The true relationship of the Moon to the Earth then came to be accepted, but still no satisfactory explanation for the origin of the Moon was forthcoming for centuries ahead.

Even in the twentieth century three competing hypotheses for the origin of the Moon persisted until the mid-1980s. These supposed, in turn: gravitational capture of an initially wandering Moon into Earth orbit, co-formation of the Moon and Earth as the solar system was formed, and fission of a proto-Earth to form the Earth and the Moon.

Consistent with ancient ideas of the 'earthly nature' of the Moon, a version of the third of these possibilities has strong support today, modern evidence of the Moon's composition suggesting that it originated in a process which has come to be known as the 'Big Splash'.[19] That event is thought to have occurred between 4.5 and 4 billion years ago, probably quite soon in cosmic terms after the solar system was formed, when, it is thought, the proto-Earth was impacted by a huge piece of cosmic debris in the form of a planetisimal about the size of Mars, that is about one tenth of the mass of the Earth itself.

There is surface evidence on Earth of impacts from cosmic debris, in the form of asteroids and meteorites that have collided with the Earth over geological time, often leaving craters as their legacy. However, it is argued that evidence of a massive impact over 4 billion years ago is to be found, not on the surface of the Earth, but partly deep in the Earth's metallic core, and partly also in the sky above. Recent cosmological and Moon-rock discoveries suggest that soon after its formation the Earth was indeed impacted by a cosmic object about the size of the planet Mars, named Thea, which threw up above the Earth a mass of rock debris and molten magma which coalesced to form nothing less than our Moon itself. Some reservations have been expressed about the likelihood of such an event, primarily whether or not a much larger impacting body would be required to make the Moon.[20] However, there is a good consensus in support of the principle of the 'Big Splash' hypothesis.[21] This was a gigantic and unique event in the sense that it led to the formation of the Moon, but impacts of cosmic debris have occurred throughout the Earth's geological history before and after the 'Big Splash'. Some geologically recent collisions have had impacts upon the Earth's biosphere, for example 65 million years ago, when a meteorite no larger than 20 kilometres in diameter impacted in the region of the Gulf of Mexico, contributing to the demise of the dinosaurs.

The Earth was, of course, lifeless at the time of the presumed 'Big Splash'. However, on that hypothesis or any other notion of how the Moon was created, present evidence strongly suggests that the Moon was created and began orbiting the Earth 4.5 to 4 billion years ago. Afterwards the Earth's crust continued to be formed, and only a few million years later, some 3.5 billion years ago, life appeared on Earth in its newly forming seas.

There is evidence from its far-side mountainous surface, contrasting with its near-side cratered surface, that the Moon was impacted by other cosmic debris after it was formed, building up its mass such that significant lunar gravitational pull would have begun to influence the Earth. In particular, as a body of about one quarter of the diameter of Earth and about one eightieth of its mass, the Moon would eventually have begun to induce substantial tidal effects on the Earth's water masses, additional to the tidal effects that would have been induced by the gravitational pull of the Sun. As the Earth's crust continued to cool, the seas were created by condensing cloud cover and probably also by water from ice that formed a constituent of incoming comets during a period of heavy bombardment 4.5 to 3.5 billion years ago. As now, daily tides induced by the pull of the Sun's gravity would be complicated and amplified by the even greater gravitational pull of the Moon, which varied according to the Earth's daily rotation and the Moon's monthly orbit. In fact, early Moon-driven tides would have occurred with greater frequency and height than at the present time. Modern estimates suggest that the daily rotation of the Earth is slowing down by about 2.5 milliseconds every hundred years and that the Moon is gradually retreating from the Earth by about 5.8 centimetres every year, thus gradually extending its current average distance of a little over 382,000 kilometres and lengthening the circumference and duration of its orbit. If these changes in the behaviour of the Moon and the Earth have been constant over time it seems likely, by hind-casting to a time around 3.5 billion years ago, that the solar 'day' and the lunar 'month' would each have been much shorter than at the present time and the Moon would have been much closer to the Earth than it is now. Accordingly the gravitational pull of the Moon would have generated more frequent and larger tides in real time, but with a similar pattern to present-day tides in the sense that there would have been two tides approximately every, much shorter, solar 'day'. It is under such conditions, with mid-ocean tides rising and falling substantially, and with frequent massive tidal excursions occurring where the sea meets the land, that early evolution of living organisms can reasonably be thought to have taken place in vast areas of tidal pools subjected to extensive cycles of wetting and drying. Even now it is the case that though tidal height in the open ocean may be less than one metre, it may exceed ten metres when amplified by the topography of the coastline.

* * *

In science, if not in some other bodies of thought, the process of the origin of life has long been a matter of speculation. However, there is reasonable scientific consensus that it began in water. Indeed there are strong arguments to suggest that life not only began in water but that its arrival was crucial in ensuring that the Earth retained its surface waters, preventing water loss that would have transformed it into a dry planet such as Venus. As James Lovelock explains in *The Ages of Gaia*,[22] the hot and dusty planet Venus lost its water when the chemical elements iron and sulphur captured oxygen from its original oceans of water, freeing up hydrogen which escaped into space. On Earth, in contrast, oxygen was generated by early life forms by the process of photosynthesis, hindering the escape of hydrogen into space and saving the Earth from 'a dusty death'. So how was this magical event brought about?

When considering the origin of life, as distinct from the origin of species, Charles Darwin speculated that it probably began in 'some warm little pond with all sorts of ammonia and phosphoric salts'. Modern views are not unlike those of Darwin, favouring the notion that life began in the early ocean which was forming as the Earth's crust was hardening during the period 3.5 to 4 billion years ago.

That ocean would have been a 'soup' in the form of a weak solution of organic molecules, the building blocks of living organisms to come. The molecules could have been formed *in situ* or may have arrived in the water contained in the impacting comets that probably provided some of the water for the primal ocean.[23] In either case the chemical elements that formed the basis of the molecules would have originated in the nuclear fission processes of dying stars which transmuted hydrogen into atoms of carbon, oxygen, and other elements in the periodic table. But by what means could basic organic molecules, so-called monomers, be induced to aggregate into more complicated molecular structures that have the capacity to self-replicate, a process which is the basis of the ability to reproduce, a vital characteristic of living organisms?

Early suggestions that life began in the depths of the sea were discounted when chemical evidence purported to show that the process of polymerization, whereby simple organic molecules are transformed into large complex ones, cannot be achieved in an aqueous medium.

Notwithstanding that argument, suggestions that life began around deep ocean hot water vents are still put forward. However, perhaps life began, not in the open sea itself, but in rock pools that fringed the primal seas. As we have seen, with closer proximity of the Moon to the Earth at that time, the twice 'daily' tides would have been more extensive than now, generating great extremes of wetting and drying in rock pools at the highest levels of a sea shore.[24] Moreover, alternate wetting and drying of high shore pools under hot sunlight would not only have occurred with the periodicity of twice 'daily' tides. Such pools would also have been subjected to an approximately twice-monthly pattern of tidal rise and fall from neaps to springs. As will be explained later, for reasons associated with the relative movements of the Moon around the Earth during their combined orbit of the Sun, neap tides are of smaller range than spring tides, with which they alternate every seven days or so (Fig. 4).

As a consequence, as at the present time, some tidal pools high on a shore at the edges of the primal seas which would have been flooded twice daily during spring tides, may not have been reached by high tides during neap tides. They may have been subject to continuous desiccation for up to several days until fully refreshed by flooding high spring tides that again surged to the highest levels of the shore around the times of full and new Moon. So it was perhaps the case that life itself began under the influence of the Moon, simple molecules in the primal soup being transformed into complex,

31

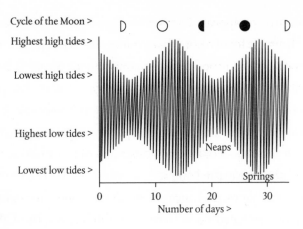

Fig. 4. Monthly cycle of the Moon, with timings and ranges of spring and neap tides.

self-replicating molecules by the wetting and drying processes under the Sun. In this way, Moon-induced tidal action, greater than that of the present day, can be envisaged as having taken the first step in establishing life forms on the fringes of the hitherto lifeless primal oceans.

An additional step that would have been required, to ensure the creation of complex cellular plants and animals, would have been for self-reproducing molecules, or DNA, to become enclosed in a protective cellular structure, a process which has been demonstrated in the laboratory. Encapsulation of DNA molecules within fatty vesicles, called liposomes, was achieved by exposing such a mixture to wetting and drying cycles mimicking those that occur in high shore tidal pools,[25,26] further supporting the hypothesis that lunar gravity was involved in the origin of life. However, whatever the process by which living organisms first appeared on Earth some 3.5 billion years ago, it is clear from the fossil record that evolution by self-replication and natural selection proceeded for many millions of years in water only. The colonization of dry land by animals and plants occurred much later, insect colonization beginning in the Silurian period of the geological record, little more than 400 million years ago. For around 90 per cent of the time that living organisms have existed on Earth, they have lived in water, many in ocean localities, the fringes of which were under the influence of Moon-generated tides. Hence lunar influences, through the effects of tides, may not only have been involved in the creation of life itself and in the maintenance of the chemical stability of Planet Earth as we know it today, but they may also have had pervasive effects on living organisms in coastal areas over aeons of time.

Throughout much of the early history of life on Earth, as the continents were formed and the great diversity of plants and animals evolved, the perpetual orbit of the Moon and its influence on tides determined similarly repeated patterns of environmental change affecting coastal animals and plants in shallow seas almost wherever they occurred. Would it be surprising, therefore, that such organisms have adapted to those lunar and tidal periodicities in the same way that organisms exposed only to the repeatability of solar day and night have acquired internal daily clocks? Such potentially impressive adaptations to the conditions of our cycling planet could not have been envisaged by Charles Darwin in 1859 when he wrote his often quoted, but apposite, lines that 'whilst this planet has gone cycling on according to the fixed laws of gravity...endless forms most beautiful and most wonderful have been, and are being, evolved'.

The most obvious environmental cycle to which Darwin referred is that of day and night and it is certainly the case that most plants and animals, including human beings, do exhibit adaptations to the light–dark cycle of the solar day. These 24-hour patterns of adaptation are reflected in so-called circadian (approximately daily) rhythms, which we become particularly aware of when our sleep and digestive patterns become disrupted by jet-lag after long distance air travel. In humans, it is well known that such rhythms are controlled by our own internal biological clocks. Most animals and plants also exhibit such biological rhythms of approximately daily periodicity, again under the influence of internal biological clocks that have evolved against the backdrop of the solar-day spin of the Earth in relation to the Sun.

But the Earth is also spinning in relation to the Moon, which is itself orbiting the Earth approximately every month, and we may ask whether evolutionary adaptations similar to those that follow the day/night cycle have occurred against the evolutionary background of what are perceived from Earth as cyclical changes in the position and size of the Moon. What are those lunar cycles and how do they compare with the perceived 24-hour solar day cycle? In the Earth's 24-hour spin in relation to the Moon it has to be remembered that the Moon is also moving in its orbit around the Earth. Consequently, to an observer on Earth, a little catch-up time is required for the Moon to be seen in the same place in the sky on successive nights. The catch-up time is slightly less than an hour every 24 hours, giving a 'lunar day' of 24.8 hours as distinct from the 24-hour solar day. So, if we want to determine whether adaptations have occurred in relation to the Moon as well as to the Sun, it is necessary to distinguish quite subtle differences in the periodicities of cyclical patterns of behaviour that an animal might exhibit. But identification of lunar cycles to which living organisms may have adapted does not stop there. We also need to consider the possibility of adaptations to the observed waxing and waning of the Moon during its monthly cycle.

Here it is important to recognize the difference between the sidereal month and the synodic or true month. The sidereal month of 27.3-day periodicity is the time taken for the Moon to complete an orbit of the Earth, during which time it returns to the same point overhead relative to the Earth. However, this is not a periodicity that is readily recognized by an observer at a fixed point on the Earth. The synodic or true month is the periodicity that is most readily perceived. Because the Earth and the Moon together move in their joint orbit around the Sun, during the Moon's sidereal orbit of the Earth, the Moon does not return to its position on a line between the Earth and the Sun for another 2.2 days. Consequently, the interval between successive full Moons as viewed from a fixed position on Earth is 29.5 days, which is the duration of the observed 'true' lunar month. This determines why there are not exactly twelve or thirteen lunar months in a calendar year, but somewhere between the two.

This is the cyclical background against which we must look for evolved responses and adaptations to the cycling of the Moon as perceived from Earth. As we will see in Chapter 3, adaptations to indirect effects of the Moon have for some time been demonstrated unequivocally through the influence of ocean tides on animals and plants that live in coastal localities that are exposed to tidal rise and fall. But how does tidal rise and fall come about?

The theory of tides was first developed by Isaac Newton (1642–1727) from his laws of universal gravitation, but the notion that the Moon causes tidal rise and fall pre-dates Newton. Johannes Kepler (1571–1630) was among the first to advance the idea that the Moon exerted a gravitational pull on the water of the ocean, attracting it towards the location on Earth where the Moon was overhead, forming a bulge as it did so. He also explained why the water did not flow skywards towards the Moon, because of the balancing effect of the pull of the Earth's gravity. It was Newton, however, who explained that though the Earth rotated beneath the Moon only once every 24 hours, there were typically two high tides during its daily rotation, not one. The explanation arises from an understanding of physics. Two tides occur during each lunar day because in addition to an upward bulge in the ocean surface attracted by gravitational pull of the Moon immediately above, there is a compensatory bulge generated in the ocean on the other side of the Earth. The two bulges balance each other and appear to travel around the Earth in tandem as it spins on its 24-hour cycle, the extra 50 minutes of the lunar day appearing, as has been seen, because of the orbital movement of the Moon, typically generating tides at 12.4-hour intervals at any one locality.[27]

Convincing though it might seem that animals and plants have adapted indirectly to the Moon through the influence of tides, it has also to be taken into account that the intervals between tides that we experience today may not always have been the same. As we have noted earlier, extrapolating backwards from present understanding that the Moon is very slowly retreating from the Earth and that the Earth's rate of rotation is very slightly slowing down, it has been estimated that tidal intervals were shorter than 12.4 hours soon after life appeared on Earth. With a faster spinning Earth and a shorter orbit of the Moon it has been estimated that some 300 million years ago, in the Palaeozoic era, tidal bulges would have progressed, not at 12.4-hour intervals, but rather more frequently at intervals of around 10.3 hours. In other words, the lunar day would have been of 20.6-hour, not 24.8-hour, duration. Consistent with these hindcasts it has been estimated, too, that during that time there would have been more than four hundred days (that is, revolutions of the Earth) during the Earth's year-long orbit of the Sun, and there is fossil evidence to support that estimate. Close analysis of the growth rings of fossil corals from some Palaeozoic rocks indicates that they possess around four hundred presumed daily rings within each annual ring.[28] So, it seems likely that such corals, and other marine organisms too at that time, were responsive to and presumably genetically adapted to the temporal regime of the environment, such that coastal species were adapted to tidal regimes of 10.3-hour periodicity. Nevertheless, the huge timescale of evolution is such that, even though the earliest organisms around marine coastlines were adapted to existing tidal patterns, their descendants would readily show genetic adaptation to the slowly increasing tidal interval over geological time until the present day.

At the outset though, despite evidence that animals and plants living in coastal seas have acquired inbuilt adaptations to tides, and therefore indirectly to the Moon, doubts are sometimes expressed about this in view of the variability of tides in relation to locality, time, and weather. Clearly the height of tides varies throughout the lunar month, and there are additional differences in the pattern of tides associated with locality and coastal topography that make it necessary for calculations used to prepare predictive Tide Tables to be specific for the locality in question. In addition there are sporadic effects on tidal range and timing attributable to wind and barometric pressure. As one expert on tidal biological rhythms has written, after considering the nature of tides:

> why would Mother Nature have even tried to develop a living clock to time an organism's behaviour and physiology to such a

messy, discombobulated, erratic cycle as the ebb and flow of the tides?[29]

However, underlying all the variability, the fundamental periodicity of tides, as half of the periodicity of the lunar day, has provided, and continues to provide, strong and recurrent environmental signals for coastal organisms over long periods of evolutionary time, despite the 'noise' and vagaries of wind and barometric pressure. Moreover, the environmental signals associated with tidal rise and fall are substantial, including as they do cycles of immersion and emersion, repeated increases of barometric pressure and reduction of light intensity as the tide rises, and potentially significant temperature changes between air and water. Over longer timescales, too, there are fortnightly changes in tidal height between neap tides and spring tides, which are again Moon-driven. Spring tides occur when, during the Moon's orbit of the Earth, the Sun, Moon, and Earth are all in line, with the Moon visible from Earth as new or full. At those times the combined gravitational pulls of the Sun and Moon produce the exaggerated tides of springs. At its quarters the Moon is at right angles to the Earth–Sun axis and the gravitational pulls of the Sun and Moon compete to produce weaker neap tides. However, the pattern of Moon-induced tides prevails, because the gravitational pull of the Sun is slightly less than half that of the Moon. Strikingly, too, the neap tide–spring tide cycle induced by the Moon induces similar cyclical changes in the rates of flow of tidal currents in coastal waters. Tidal currents are variable according to locality but typical springs tidal currents of a few knots (nautical miles per hour) are two to three times greater than those of neaps. It is, therefore, difficult to avoid the conclusion that many organisms on Earth that are repeatedly exposed to such cyclical environmental events may reasonably be expected to have shown evolutionary adaptations, indirectly, to the lunar cycles that induce these rhythmic physical phenomena.

More questionable is the possibility that organisms have adapted directly to lunar cycles such as changes in the intensity of moonlight. Variations in the amount of moonlight between day and night are masked by the intervention of daylight and, moreover, differences of light intensity between new and full Moons are relatively small, which suggests to sceptics that lunar cycles may have been of little direct evolutionary significance to plants and animals, including humans. Anecdotal accounts of Moon-related phenomena are often scorned in the present day, perhaps not surprisingly in view of the lunar myths and legends that have prevailed throughout human history. However, despite such scepticism, there is some evidence of direct effects of the Moon on living organisms, evidence that will be presented later in this book. Moreover, modern molecular biological investigations are beginning to provide evidence for inherited biological clocks of both tidal and lunar periodicities in some animals,[30,31,125] leading to the conclusion that heritable adaptations to lunar cycles have occurred during evolution. Throughout the book, I shall be seeking to show, by scientific methods, indirect and direct adaptations to the Moon, an approach contrasting both with ancient myths and legends and with later anecdotal accounts of such associations which may or may not have validity.

3

THE MOON, THE UNICORN, AND TIDAL MEMORIES

Just as there is a history of popular conceptions of supposed relationships between lunar cycles and the life of humans, so too there is a history in science of recorded relationships between the Moon and the behaviour of animals and plants. Some of these relationships have been shown to be real, with scientifically valid explanations, but many too are 'supposed' and rightly viewed with scepticism. The problem, then, is that when some so-called Moon-related phenomena are shown to be spurious it becomes easy for sceptics to insist that all such phenomena are equally questionable. Even among scientists in recent decades, the notion of lunar myths and legends looms large and, intuitively, the concept that any animal or plant may be responsive to the cycling Moon has taken time to become acceptable. It is hoped that this and following chapters will remove any lingering doubts by providing scientifically estab- lished evidence that some plants, and particularly animals, have been shown, unequivocally, to exhibit Moon-related behaviour, indirectly through the influence of tides and even directly in relation to lunar cycles.

* * *

As early as the closing years of the nineteenth century it was claimed by naturalists that one of the brown seaweeds (*Dictyota dichotoma*), commonly occurring on the coasts of Europe, reached sexual maturity in a cyclical pattern related to the Moon.[32,33] The pattern was seen to be fortnightly with peaks of reproduction occurring during spring tides, more or less coincident with the times of new and full Moon. This early discovery was, unsurprisingly, initially regarded with scepticism, but it proved not to be merely anecdotal. It was quickly demonstrated that the same Moon-related rhythm of reproduction persisted in seaweed brought into a laboratory and kept in aquaria away from the influence of the moonlight and tides. This suggested that the reproductive rhythm was inbuilt, ensuring wide dispersal and survival of the species by distributing reproductive bodies in strong currents associated with spring tides. If there were lingering doubts about the validity of this Moon-related phenomenon in a seaweed they were addressed sixty years later by a German botanist, Dietrich Muller, who first confirmed that *Dictyota* did indeed persist with a fortnightly rhythm of reproduction in the laboratory.[34] Crucially then, noting that the breeding pattern became random after some days in the laboratory, he was able to reinstate the rhythm, still in the laboratory, by exposing the seaweed to artificial cycles of moonlight at night to simulate waxing and waning of the Moon.

* * *

By the time in the mid-twentieth century that Muller confirmed the occurrence of a Moon-related rhythm in a seaweed, a new science was emerging concerning adaptations of animals and plants to the cyclical nature of their environment. It was at that time that the notion of biological clocks began to be formulated. This arose from the clear demonstration that plants and animals, including humans, are able to maintain bodily rhythms and behaviour patterns, virtually in phase with the outside environment, for some time after being placed in complete isolation from daily changes of sunlight and temperature. The concept of circadian (approximately daily) rhythms controlled by internal biological timekeeping systems was formulated on the basis of such observations, a concept that is now well established in the public domain.[35]

Less well accepted were more recent findings that animals from the seashore, when removed from the influence of Moon-driven tides and transferred to controlled conditions in the laboratory, express bodily rhythms and behaviour patterns that clearly reflect the approximate timing of tides on the shores from which they were removed.[27] Shore crabs (*Carcinus maenas*), which are well known to secrete themselves in rock crevices at low tide and forage at high tide in a sleep–waking pattern, continue with that pattern when brought into the laboratory. For a few days in constant conditions, freshly caught crabs consistently signal the time of high tide on the shore from which they came, by periodic increases in their walking behaviour and exploratory movements. By analogy with circadian rhythms, these patterns of behaviour are now defined as circatidal rhythms, initial circumstantial evidence suggesting that they too are driven by body clocks, not of approximately daily, but of approximately tidal periodicity. Circadian clockwork has for some time been known to have a molecular basis, suggesting that changes of behaviour relating to apparent movements of the Sun are deeply engrained in the genetic make-up of plants and animals and, therefore, are an inherited feature of living organisms. A question has persisted, however, as to how the timing of circatidal rhythms is maintained. Are they, as approximately 12.4-hour rhythms, controlled by circadian clocks operating in antiphase, or are they controlled by dedicated circatidal molecular clocks? Only recently, in 2013, has this debate been advanced significantly, with evidence for molecular clockwork of approximately tidal periodicity, quite distinct from clocks of circadian periodicity,[30] to be discussed later in the book. However, the study of biological rhythms, particularly tidal and lunar rhythms in marine organisms, has had a long history of controversy and it has often been argued that they are not controlled by internal biological clocks at all. Argument persisted for some time that when organisms were brought into so-called constant conditions, no attempt was made to control 'residual' environmental cycles of factors related to variables such as the Earth's magnetic field, barometric pressure, and cosmic ray bombardment. Some scientists suggested that one or more of such factors might be an external timer for behaviour that only appears to be controlled by internal clockwork.

Internal clockwork has been confirmed, but a salutary tale still circulating on the Internet, long after it was first told in 1954, concerns a study of the respiration rhythms of the humble oyster.[36] That study purported to show that oysters transported from the east coast of the USA to an inland laboratory in Illinois re-set the timing of their tidal rhythms of shell opening and closure to correspond with times of the Moon's zenith in their new locality. It was claimed that the oysters were responding directly to daily changes in the Moon's gravity, a claim being accepted by surfers of the Internet some sixty years after the claim was first made. In fact, quite soon after the original claim for direct perception of changes in lunar gravity was made it was pointed out in the scientific literature that if the findings were justified they would represent an amazing scientific breakthrough concerning the sensory capabilities of oysters, or indeed of any living organism.[37]

A particular problem concerning the original claim was that in analysing long series of data the statistical methods used often gave dubious results. As a result, when other scientists came forward to check the original findings, they questioned whether the oysters showed any such Moon-related rhythms at all. Indeed, in 1957, an eminent American biologist, Lamont Cole, writing in the distinguished journal *Science*, successfully spread scepticism among biologists in general concerning the notion of Moon-related biological rhythms by demonstrating a biological rhythm in the mythical Unicorn.[38] What better creature could be chosen for such a study, in view of the fact that the Unicorn is thought by some historians to have been a calendar symbol in ancient Greek mythology? As a composite animal, consisting of the body of a horse (with a central twisted horn), the hind legs of an antelope, and the tail of a lion, its tripartite structure represented the ancient three-season year, somewhat analogous to the Greek triple Moon-goddesses symbolizing the new, full, and waning Moon. Having selected such an appropriate animal for the study, the only task then was to quantify some aspect of the behaviour of the Unicorn that varied over time, to provide data which could then be analysed to determine whether the creature exhibited some form of biological rhythmicity. For an imaginary animal this was easy, and the study began by generating a sequence of random numbers that were assumed to be imaginary hourly values of oxygen consumption throughout a sequence of a few days. Amazingly to some, but not to Lamont Cole, it was revealed that the Unicorn had a clear daily pattern of 'respiration', breathing particularly energetically at regular intervals. Coincidentally the peaks of 'respiration' occurred during the assumed hours of darkness in the long data set of imaginary hourly values of oxygen consumption that were used in the computation, from which it might be concluded that the Unicorn is nocturnal!

So how was this 'nocturnal' pattern of behaviour achieved by a mythical animal that could not possibly possess an internal biological clock and which could not be influenced by external environmental factors? The answer to the question was known immediately to Cole, who, in his analysis of the sequence of hourly values of 'respiration' rates, had set out to use statistical procedures that were in common use at the time and which were known to the author to have faults. Such procedures for the analysis of time series of data were known to be questionable when seeking to distinguish between Moon-related and Sun-related patterns of rhythmicity. Moreover, it came to be recognized that there were additional faults with the statistical procedures in use at the time when it became apparent that the methodology itself induced rhythmic patterns in the analyses of randomly generated data.

Therefore, as a critique of studies which sought to show that oysters varied their rate of respiration directly according to the phases of the Moon, an article entitled 'Biological clock in the Unicorn', in a respected scientific journal, certainly heightened scientific scepticism concerning Moon-related behaviour in living organisms. Nowadays such an article might appear in *The Annals of Improbable Research* and qualify the author to compete for the award of the Ig Nobel Prize for 'science that makes you laugh and then makes you think'. Appearing in *Science*, which gave it strong credence in the scientific community, it raised even higher than before the level of scepticism concerning the notion of Moon-related rhythms in animals and plants. The article was surely grist to the mill of anyone denying relationships between the Moon and life on Earth, particularly when viewed alongside the range of pervasive and unjustified beliefs, myths, anecdotes, and perceptions from the earliest times of human awareness of the natural environment. Generalized rejection of the notion, based upon selective evidence of this kind, has ensured that, over the years, there has developed considerable rational scepticism concerning the reality of any influence of the Moon on living organisms, including man. So much so, that the respectability of scientific study of Moon-related phenomena in the natural world has sometimes been questioned. How then are supposed lunar influences on the behaviour of living organisms being brought into the scientific domain?

To achieve the required scientific rigour it is first necessary to demonstrate consistent and repeatable correlations between the Moon or tide and the behaviour in question. Then some way should be found to determine whether the behaviour occurs in isolation from Moon or tide, and whether its timing can be manipulated by artificial cycles of lunar or tidal periodicity. If so, then true adaptation to Moon or tide can be considered to have occurred and that an organism capable of behaving in this way is in possession of lunar or tidal biological clockwork. Just as it is clear that our own circadian body clocks can be re-phased by new environmental cycles of solar periodicity, when we carry out the experiment of jetting between time-zones,[35] we shall see that circalunar and circatidal body clocks can be similarly re-phased, by new environmental cycles of lunar and tidal periodicity. The biological rhythms of various coastal animals have been shown to be responsive to simulated cycles of moonlight, water pressure, wave action, and temperature.[27] Finally, after establishing that lunar and tidal influences are so significant that biological clocks have evolved to match those influences, consideration has to be given to the adaptive advantage that the clocks endow. For example, a beach-living animal that 'preferred' to live in water might be thought to find it of evolutionary advantage to prepare for occasions when it was left high and dry by the falling tide. Equally, one that 'preferred' to live in air, but found other advantages in living on a tidal shore, might be thought to find it advantageous to prepare for the regular incursions of sea water. Body clocks of circatidal periodicity would enable them to achieve such timings. Similarly, body clocks of circalunar periodicity could enable animals to distinguish between the strong tides of springs and the weaker tides of neaps if it was a disadvantage to them, for example, to be carried offshore in strong tidal currents during spring tides.

* * *

One of the first truly scientific accounts of seashore animals that are able to anticipate the ebb and flow of tides was reported by the scientists F. W. Gamble and F. Keeble studying beaches of western France near Roscoff, Brittany around 1900.[39] They noted that, during daytime, patches of green coloration appeared on the beach surface as the tide ebbed, only to disappear again before the tide returned. The green colour indicated the emergence on to the beach surface of large numbers of small flatworms, *Convoluta roscoffensis*, each containing single-celled green plants, or zooxanthellae, that live symbiotically in the worm tissues. The animal/plant association was found to be a true symbiosis, the worms transporting zooxanthellae on to the sand surface during daytime low tides, where they were able to use light for photosynthesis, and ingesting as nutrients the waste products of zooxanthellae metabolism. However, fascinating as it would have been at the time to have discovered such an intimate symbiotic association, it was surely remarkable for the scientists to discover that the flatworms did not wait to be covered by the rising tide but were able to re-burrow in the sand ahead of the time they would expect to be covered by flooding sea water. Apparently the worms were able to anticipate the time of the rising tide, ensuring that, by burrowing, they avoided being washed away from their optimal zone on the beach just above mid-tide level.

The importance of their discovery was not lost on Gamble and Keeble, or on other marine biologists, who immediately began to ask questions about how the flatworms were able to regulate their behaviour in such a way as to enable them to anticipate the rise and fall of Moon-driven ocean tides. Was it possible that the burrowing worms were simply responding to vibrations of waves during the advancing tide, or did they possess an internal sense of time? Moreover, if they did possess an internal sense of time, was it a simple timing sense equivalent to an egg-timer that measured off the length of time that the worms were exposed to the air from the time of first exposure during the ebb? Or was it a more complicated sense of timing than that? Critically, though, the scientists who discovered the burrowing pattern of Convoluta not only asked themselves some of these questions but also sought ways in which to find out answers to their questions.

Accordingly, in the year 1903 when the first report was published describing the behaviour of Convoluta on the beach, another scientist, G. Bohn, aware of that behaviour, published an account of experiments on the behaviour of the flatworm under experimental conditions in the laboratory.[40] In order to find out whether the flatworms were responding to the rise and fall of tides, freshly collected animals were placed into sand in laboratory aquaria which were supplied with a constant flow of sea water and completely isolated from the effects of tides. In this simple experiment it was shown that the flatworms continued with their spontaneous pattern of emergence on to the sand surface at the approximate times of low tide on the shore outside, re-burrowing again before the time of high tide, exactly as they would have behaved had they remained in their natural habitat, even though they were completely immersed in sea water throughout the experiments. Intriguingly the same pattern of emergence and re-burrowing continued as a tidal rhythm, more or less in phase with tides on the shore outside, throughout a period of four days, with no possibility at any time that the flatworms could have responded to tidal rise and fall. After about four days in isolation in the laboratory the flatworms continued to emerge on to the surface of the sand in the aquarium tanks, but did so in a random manner with no synchronization to the timing of tides. It was as if the flatworms possessed some form of biological clockwork of approximately tidal periodicity that became de-synchronized in the absence of clock-setting signals that they would normally receive from Moon-driven tides in their natural environment. And this was indeed confirmed in additional experiments when flatworms that had apparently lost their rhythmicity after several days in the laboratory were returned to the beach to assess whether their tidal rhythms could be reinstated. In those experiments flatworms were returned to the beach in containers from which they could not escape yet were again exposed to the rise and fall of tides. After a few days back in their home environment the flatworms were again tested in the laboratory aquaria and found to have re-acquired the ability to express tide-related rhythms of emergence when kept in constant conditions.

These experiments suggested, for the first time, that an animal, even as simple an animal as a flatworm, possessed a tidal memory, that is an internal clock mechanism, which permitted it to measure time in relation to the timing of Moon-driven tides. These findings were unusual at the time, no doubt prompting a further study a few years later by G. Bohn and H. Pieron,[41] and another by L Martin,[42] all of which confirmed the earlier studies, emphasizing these novel findings by the use of such terms as 'la mémoire chez *Convoluta*' and 'la phénomène de l'anticipation réflexe'.

Subsequently, many other mobile organisms living between tidemarks around the world's oceans, from single-celled plants, through worms, molluscs, and crustaceans, to fish, have been demonstrated by similar experiments to have the ability to show tidal anticipatory behaviour.[27,29] Sessile animals such as barnacles and mussels that are permanently attached to rock surfaces seem to survive without anticipating tidal rise and fall. They simply open and close their shells in response to wetting and drying due to tides, but motile organisms, for which there would clearly be advantages in anticipating tidal rise and fall, appear to have evolved tidal biological clockwork to permit them, say, to seek shelter in sand or in rock crevices before being left high and dry by the falling tide.

Though most biologists would now accept the notion of internal tidal clockwork in coastal organisms, not all investigators have agreed with that concept, particularly in the early years of the new science of chronobiology. As we have seen, in the past there has sometimes been a reluctance to accept the possibility that internal clocks control tidal rhythms, leading some investigators to postulate that when animals are maintained in the laboratory in so-called constant conditions, they might still be influenced by environmental variables that are not under the investigator's control. In other words, so-called 'residual periodic variables' such as barometric pressure changes in the atmosphere and even changes in the pattern of incoming cosmic radiation might be the triggers for observed rhythms of behaviour in organisms maintained in supposed isolation away from their normal environment.[43] However, no evidence supporting such sceptical views has been forthcoming and, in any case, if organisms are capable of using such residual environmental cues to time their biological clocks, why do they not continue to utilize those cues in perpetuity when they are brought into the laboratory? Biological rhythms in the laboratory, of whatever periodicity, become randomized after some days, as though they are controlled by internal clockwork that is deprived of more obvious external environmental synchronization. The generally accepted view now is that many coastal organisms are indeed able to anticipate environmental changes that are associated with Moon-driven tides but are reliant on repeated exposure to clearly apparent Moon-related environmental variables to keep them in time.

* * *

Let us now take the example of the sea louse, an isopod crustacean related to garden-living woodlice, which at low tide can be found buried in sand to a depth of a few centimetres just below high water mark on many European beaches. It clearly shows tide-related swimming behaviour,[27,44] and human swimmers sometimes find themselves 'stung'—or more correctly 'bitten'—by the sea lice which often swim in large numbers at high tide. Several species of sea louse occur on beaches

Fig. 5. *Eurydice pulchra*— a sea louse.

throughout the world but on many British and European beaches the commonest species is *Eurydice pulchra* (Fig. 5).

Like the Greek mythological character after which it is named, *Eurydice* spends part of its time buried in the under-world, in this case buried in the sand of the beach. But this *Eurydice* is not reliant on Orpheus to be led from the under-world; males and females emerge together from the sand to swim in the water above, there to feed and find a mate, before returning to burrow in the sand and remain hidden during low tide. The zone in which they are found burrowed in the beach at low tide is consistently high on the shore. The location of the animals bears a constant relationship to the level reached by the tide, in such a way that their zone of occurrence moves up and down the beach during the fortnightly cycle of neap–spring tides. Exactly the same behaviour occurs in a related species of sea louse, *Excirolana chiltoni*, on beaches in California.[45] In both species this is an intriguing pattern of behaviour, reminiscent of the behaviour of the flatworm *Convoluta* described earlier in this chapter, and a key question relating to that behaviour is essentially the same as the one discussed in relation to the flatworm. In *Eurydice*, for example, it can be readily understood that the animals might emerge to swim when they are covered by the rising tide, but how do they determine when to burrow in the sand in order to avoid being swept out to sea when the tide turns? How does *Eurydice pulchra* maintain its preferred position buried in the sand in the upper part of the beach and so avoid being carried to unsuitable habitats offshore where they would be in unfavourable competition with related sea lice that are specially adapted to live there? The answer is that, like *Convoluta*, they possess their own internal biological clocks that oscillate at approximately Moon-driven tidal periodicity. The clocks are re-set by wave action during the rising tide, which also encourages the sea lice to emerge and swim, after which, importantly, the clocks determine the time when the sea lice re-burrow well in advance of low tide, by which time they might otherwise have been carried offshore.[27]

As a resilient species which in normal life tolerates being buried in sand throughout the low tide period, *Eurydice* also survives well in laboratory aquaria. Its behaviour has therefore been studied in detail in the laboratory by biologists who have shown that it exhibits a tide-related pattern of swimming even when kept in controlled conditions away from the influence of tides. Even in aquaria, freshly collected, laboratory-maintained specimens emerge quite spontaneously to swim at the approximate times of high tide that they would experience if they had remained on their native beach, re-burrowing again when they expect low tides, confirming that they possess inbuilt biological clocks that run at approximately tidal periodicity. More than that, however, they also exhibit a monthly tidal pattern of spontaneous swimming superimposed upon the twice-daily tidal pattern of swimming. Spectacularly, it is the case that when a collection of *Eurydice* is kept for a long period of time in laboratory aquaria, with or without sand at the bottom of the tank in which to burrow, greatest numbers swim quite spontaneously during those time of the month when they would have expected spring tides to occur if they had remained on the beach from which they were collected. In short, greatest numbers can be seen swimming under their own volition just after the times of full and new Moon, without seeing the Moon and with no experience of tides. What better evidence could there be of animals not only showing behaviour related to lunar cycles, but also of having adapted to such cycles to the extent that the Moon-related behaviour is built into their genetic make-up? Later in the book we will consider evidence from molecular studies for clock genes that control tidal and lunar patterns of behaviour in *Eurydice* and in a marine worm respectively. Moon-related behaviour in living organisms might not, then, always be mythical but in some cases have been achieved by evolutionary adaptation.

What, though, is the advantage to the sea louse of swimming most abundantly just after the times of new and full Moon? These times, of course, coincide with the fortnightly cycle of spring tides, so that with the combination of their daily tidal and their lunar monthly rhythms of swimming they swim in greatest numbers after the high waters of spring tides. In doing so the population would tend to be carried a little way down the shore and would avoid being stranded on their native beach at levels that would not normally be reached by high tides during intervening neap tides. In other words, their fortnightly, semilunar, rhythm of swimming ensures that *Eurydice* avoids 'neaping', an event familiar to boat owners when their craft are moored near high water mark and may not be reached by incoming tides during periods of several days around times of the quarter Moons.

So we have identified Moon-related behaviour, timed by an internal clock biological mechanism of fortnightly periodicity, which is apparently beneficial for the survival of sea lice in their preferred habitat between tidemarks on sandy beaches. The Moon-related biological clock, like the tidally related biological clock, may have a molecular, and therefore a genetic basis, but what is not clear is how the Moon-related clock is synchronized. Since increased swimming occurs during spring tides, just after new and full Moon, it seems unlikely that perception of moonlight itself is the direct synchronizing mechanism of the fortnightly clockwork. Moreover it seems even less likely that this is so in view of the fact that *Eurydice pulchra* is a native of beaches of northern Europe where, because of weather patterns, moonlight is an unpredictable clock synchronizer. It is, however, much more likely that the Moon is an indirect synchronizer through its gravitational influence on tides.

This turns out to be the case, and it does not have to be taken on trust. In simple experiments with an aquarium stock of *Eurydice* which over time lose their spontaneous rhythmicity, a semilunar rhythm can easily be reinstated. By generating artificial wave action to simulate agitation by the waves of high spring tides at particular times of day, the fortnightly rhythm of swimming can be completely re-set (Fig. 6).

Intriguingly, as is often the case in scientific research, finding answers to one set of questions generates new questions. Since spring high tides occur at different times of the day in different geographical localities, covering the geographical range of distribution of *Eurydice*, we can study the synchronization of Moon-related rhythms in *Eurydice* from localities with different tidal regimes, that is in which highest tides occur at different times of the day and night. For example, because of the topography of the coastline of the British Isles, pulses of high tide

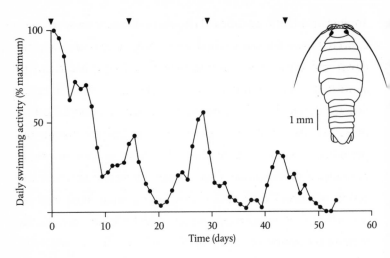

Fig. 6. Entrainment of a semilunar (approximately 14-day) free-running rhythm of swimming activity in a laboratory stock of the sea louse (*Eurydice*) maintained in constant conditions after exposure to artificial tides for four days.

take time to travel northwards through the Bristol Channel and Irish Sea such that high (and low) tides occur about six hours later on the north coasts of Wales than on the south coasts of that country. Relatively easily, therefore, it was possible to compare the timing of the biological clocks of sea lice collected in localities where high spring tides occur around midday and midnight with the timing of the clocks of others collected where maximum tides occur around six hours later at dawn and dusk. In experiments carried out concurrently, it transpired that sea lice collected at a locality where high spring tides occur around midday and midnight were able to time their semilunar rhythms in response to artificial tidal agitation applied around those times of day. In contrast, with specimens collected from a locality where high spring tides occur around dawn and dusk, it was found necessary to expose them to simulated wave action at those times in order for their fortnightly clockwork to be re-set. So we can conclude that these sea lice not only show genetic adaptation to the periodicity of twice-daily tides, but also to the monthly periodicity of the lunar cycle, albeit indirectly through the action of spring tides. Even more intriguingly it is evident, too, that local populations of sea lice have adapted to the fortnightly timing of spring tides in relation to the solar day, which varies according to their geographical position.[46]

So, we know that sea lice possess internal biological clocks of tidal and fortnightly periodicities that have evolved indirectly in relation to lunar cycles, and there are indications that these clocks interact in some way with the day/night cycle. Hence we can now ask whether it is possible that these creatures also have internal daily clockwork that has evolved in relation to the solar day, and the answer to that question is yes. The pulchritudinous appearance of the animals which gives them their name of *Eurydice pulchra* is determined by a beautiful array of black colour cells, or chromatophores, which expand and contract in a pattern that matches the solar day. The chromatophores expand to darken the surface of the animals by day and contract to result in a paler appearance by night, in a timed sequence of events that persists even in animals kept in constant dim light, suggesting that the sequence is under the control of internal clockwork of approximately 24-hour, that is circadian, periodicity. It can be said with confidence, therefore, that the sea louse *Eurydice pulchra* possesses at least three types of biological clock—circatidal, circasemilunar, and circadian—each of which is probably under gene control. Since it is widely accepted that clock genes of circadian periodicity, first discovered in invertebrate animals, are widespread throughout living organisms, including Man, it has to be considered that circatidal, circasemilunar, and perhaps even circalunar clock genes may be similarly widespread. John Steinbeck certainly seemed to concede this possibility when he wrote, in *Cannery Row* in 1945:

> He (Doc, a Marine Biologist) didn't need a clock. He had been working in a tidal pattern so long that he could feel a tide change in his sleep.

In later chapters we shall now consider the extent to which a wide variety of animals are able to feel a tide change or even a Moon-change, 'in their sleep'.

4

ARISTOTLE'S URCHINS AND DANCING WORMS

In classical Greek culture, and probably earlier, among coastal-dwelling peoples who were partially dependent on the sea for food, there was a belief that the quality of edible shellfish varied throughout the lunar month. This awareness was expressed by Aristotle, who stated that the roes of some sea urchins, probably the edible *Echinus esculentus*, were largest during the days of full Moon, times of the month when they were said to be most favourable for eating. Here was an item of folklore that could be scorned by sceptics or put to the test by a scientist who was prepared to take it seriously and accept Aristotle's observations as the basis for a hypothesis that could be tested. Over two thousand years after Aristotle, such a scientist came forward who was prepared to test a hypothesis based on folklore of such long standing by undertaking careful microscopical investigation of sea urchin roes collected at precisely determined times of the lunar month. That painstaking work was carried out in the 1920s by Hector Munro Fox, later Professor of Zoology in the University of Birmingham, UK. In doing so, and by publishing a book on his findings, perhaps not surprisingly he may have risked the scorn of sceptical scientific colleagues.[47] However, not content with risking his scientific reputation, Munro Fox also inadvertently found himself confronting prudish members of society at the same time. He summarized his findings and his wider interest in Moon-related breeding patterns in living animals in a book published under the title *Selene, or Sex and the Moon* as a homage to the ancient lunar goddess of that name. Unfortunately, as often happens, the book was read by some for its title and key words only, and not for its content, so that it found its way to the Adult books sections of some libraries.[48]

Accordingly the book's contents not only disappointed the furtive reader, but they were also denied to the browsers of scientific books, who were deprived of a learned account of the life-cycles of shellfish by their reluctance to be seen reaching to the top shelf. In fact the book summarized a sound study of sea urchins which were collected regularly throughout a lunar month during their breeding season. They were freshly collected day by day and comparisons were made of the size and condition of their roes throughout the month. As Aristotle's observations predicted, the proportion of urchins with mature roes was shown to vary in relation to the lunar cycle. The ripeness of the roes, and hence their gastronomic quality, was shown to be best at times of full Moon, the phase of the Moon when the urchins spawned. So, the first question was answered and deniers of Moon-related biological phenomena were suitably challenged. However, as mentioned earlier, as soon as one scientific question is answered others present themselves for consideration. The advantage to sea urchins of shedding eggs and sperm *en masse* all at the same time is reasonable to understand, since synchronized spawning increases the chances of fertilization of eggs. Fertilization and development of free-swimming larvae occur externally in the sea before the larvae eventually metamorphose and settle on the sea-bed again as miniature adults, thus completing the life-cycle. But the biological advantages of breeding at a particular phase of the Moon and the cues involved in such precise timing of spawning remain unclear.

At first sight a logical explanation for Moon-phased spawning would be that it is in direct response to some aspect of the changes in tidal height that occur throughout the lunar month. Specifically in this case it might be thought to occur in response to water pressures induced by high spring tides that occur immediately after times of full Moon. However, Munro Fox's urchins (*Diadema setosum*) were residents of the Red Sea and Aristotle's experience would primarily have been of urchins living on Greek coasts, both localities where tides are very small in amplitude. In yet another locality too, in Doubtful Sound in New Zealand, another region of moderate tides, a species of sea urchin has more recently been reported to spawn only during times of the full Moon.[49] In any of these localities, it seems unlikely that the differences in tidal height between full Moon high spring tides and quarter Moon neap tides would be sufficiently great to trigger spawning. Moreover, if high spring tides at full Moon trigger spawning then why should not the high spring tides at new Moon phases of the lunar month be equally effective as a cue for spawning? For some localities at least, it therefore seems unlikely that some aspect of the tidal cycle is responsible for spawning just after the times of full Moon. So, is the intensity of moonlight operating as a cue for the synchronization of spawning? It seems unlikely that sea urchin sperm swim towards floating eggs by the light of the Moon, but it is possible that adult urchins in a spawning state are sensitive to such light and that bright moonlight itself is directly responsible as a trigger for the spawning process.

Before accepting the possibility that moonlight might be the direct stimulus for Moon-phased spawning in at least some sea urchins, a further factor relating to the nature of tides has to be excluded: the height of spring tides varies between one set of spring tides at full Moon and the next set during the following occurrence of new Moon, sometimes the former and sometimes the latter being the larger. This raises the possibility that spawning during full Moon spring tides and not during new Moon spring tides happens because full Moon spring tides are of greater amplitude than the previous and subsequent new Moon spring tides. Fortunately it is possible to answer that question because it is known that, in any one locality, the

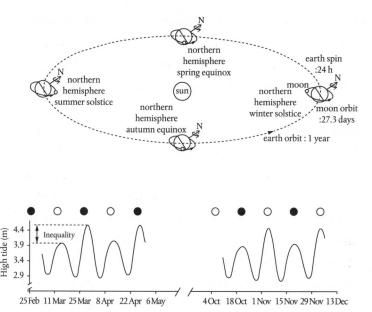

Fig. 7. Syzygy inequality cycle (SIC). Upper figure illustrates representative times of the year when the Earth, Sun, and Moon are in alignment (syzygy), generating spring tides at new and full Moon. The syzygy cycle interacts with the cycle of the Moon's proximity during its tilted and elliptical orbit from perigee to apogee to generate the 14-month syzygy inequality cycle during which maximum spring tides switch between new and full Moon every seven months (lower figure).

highest spring tides alternate between the full and new phases of the Moon on a fourteen-month cycle. Accordingly, for seven months of the year full Moon spring tides exceed those of the new Moon spring tides and then for the following seven months the opposite is the case (Fig. 7).

This pattern arises from the interplay of the cycle of the true lunar month, from one full Moon to the next, and the slightly shorter cycle of the Moon's proximity during its approximately monthly elliptical orbit of the Earth.

This pattern of tidal variation, known as the awkwardly named *syzygy inequality cycle* (see Chapter 6), prompted a critical study to be carried out on sea urchins around Santa Catalina Island, California.[50] Comparisons were made of the spawning times of a species of sea urchin during two successive summer breeding seasons, one in which the more extreme spring tides occurred during the new Moon phase and the other when extreme spring tides occurred during the full Moon phase. A preliminary study of that particular species of sea urchin (*Centrostephanus coronatus*) on the coasts of California had shown that they spawned, not during spring tides around the time of full Moon but following neap tides that coincided with times of third lunar quarters. But the question was still valid: would these urchins change their spawning times from just after neap tides following *third* lunar quarters to just after neaps following *first* quarters of the Moon during the times when extreme tides switched between new and full Moon? The answer was found to be that they did not switch: they spawned consistently just after neap tides during the *third* quarter of the Moon in both years of the study. In other words, they consistently spawned about seven days after the time of full Moon whichever tidal regime prevailed in the year of study. Fairly conclusively, therefore, it seems that spawning of the Santa Catalina Island urchins was not phased by a particular state of tide but was synchronized by some other Moon-related factor such as monthly changes in moonlight intensity.

So, Mediterranean and Californian sea urchins may spawn in response to the lunar cycle, the former releasing eggs and sperm immediately around the times of full Moon, the latter doing so with a delay of a week or so after full Moon. A new question then arises as to the significance of the differences in timing of spawning between the two localities. As will be apparent in other organisms we will look at later, a possible clue lies in the fact that full Moon periods coincide with extensive spring tides and quarter Moon periods with less extensive neap tides each month. These facts, taken in conjunction with the knowledge that tides in general would be smaller in range in the Mediterranean than on the Californian coast, perhaps offer an explanation. Release of swimming larvae into the weak tidal currents of neap tides, rather than into strong currents associated with spring tides, may be advantageous to the Californian species in not over-dispersing the larvae away from the parental locality on Santa Catalina Island, but still permitting some colonization of new localities. In contrast, in weak tidal localities in the Mediterranean the balance between over-dispersion and advantageous spread of the species is probably best achieved by release of larvae into the tidal currents induced by spring tides.

Needless to say, these observations that some sea urchins around the world spawn at particular times of the lunar month do not permit us to assume that all sea urchins everywhere show monthly rhythms of spawning. Nor is it possible to conclude that all sea urchins that do spawn at particular times of the month are cued to do so directly by the light of the Moon. Indeed some populations are known to spawn twice each lunar month, once around the time of full Moon and again at new Moon,[51] in which case they may be cued by the fortnightly cycle of spring tides, that is, only indirectly by the lunar cycle. Also, just to emphasize the variability in living systems, it has been found that, within an individual species of sea urchin that occurs over a wide geographical range, Moon-related spawning may occur at different times throughout the lunar cycle, or not at all, depending upon the geographical origin of the urchins concerned. As we have seen, however, it can be concluded from scientific evidence that some populations of sea urchins are able to synchronize their spawning to particular phases of the month. It can also reasonably be concluded that the timing of spawning periodicity in some sea urchins is somehow cued by the changing pattern of moonlight, though not necessarily moonlight intensity at a particular time of the lunar month. As we noted, the advantages of such synchronized spawning are obvious, but the phenomenon raises new questions as to how the urchins perceive changes in moonlight intensity and what role those perceptions play in the expression of the cyclical patterns of reproduction implied by Aristotle's original observations. In any case it is clear that the effect of the Moon on some living creatures is not a myth, so we are justified in asking whether the phenomenon is more widespread than originally proposed in the literature of ancient Greece.

Certainly, consistent with Aristotle's observation and wider ancient folklore of maritime peoples until the present day, it has been considered that the quality of shellfish more generally than sea urchins may vary according to the cycle of the lunar month, and scientific investigations have in some cases confirmed this. For example, in some localities king scallops have been recorded as spawning during or just after times of full or new Moon, a pattern which is entirely consistent with their 'condition', and therefore their edible quality, being optimal as they prepare to spawn during the lunar quarters. Similarly it has long been suggested that the edible quality of oysters varies according to the phase of the Moon when they were harvested. Again, changes in quality can be related to the pattern of oyster spawning, but in this case eggs and sperm are not released directly into the sea during spawning. They are retained for some time within the mantle cavity of oysters, where fertilization takes place before the fertilized eggs are released into the sea as larvae during what can more accurately be called 'swarming'.

In the mid-twentieth century the possibility of Moon-related swarming of oysters was taken sufficiently seriously by Dutch fishery scientists for them to undertake an intensive study of oyster breeding in local estuaries.[52] The biologists painstakingly counted the numbers of free-swimming oyster larvae captured by repeatedly tow-netting through standard volumes of sea water above commercial oyster beds at the mouths of Dutch rivers. Samples were obtained and the numbers of larvae counted each day throughout the oyster spawning season, from mid-June to late August, each year for ten years. Remarkably, during each year, a short but massive episode of larval swarming occurred during a 'window' of expectation from 26 June to 10 July, usually about ten days after the day of the full or new Moon that occurred during the two-week spawning window.

So, *swarming* was in some way timed to the lunar cycle, but the question remained as to when *spawning* occurred and this question was answered by studying adult oysters themselves to establish that eggs and sperm, quickly followed by fertilized eggs and larvae, appeared in the parental mantle cavities two days or so after full or new Moon. Larvae therefore spent their first week or so brooded in the parental gill chamber before they were released into the sea at the swarming event. With the times of spawning occurring around or just after times of full or new Moon, the oysters would be 'spent' at those times, so their condition and edible quality would be poor. Thus, as with the king scallops referred to earlier, during the breeding season the best-quality oysters for human consumption would be those collected at or just after times of the lunar quarters when the build-up of roes was taking place prior to spawning about a week later and swarming another ten days after that. Understandably the Dutch scientists involved in the study excluded the possibility that spawning and swarming were timed directly in phase with the Moon. After all, these events occurred a day or two after either the full or new Moon, the common occurrence at those times being the exaggerated rise and fall of spring tides in the estuaries concerned. The scientists preferred to suggest that spawning and therefore subsequent swarming were triggered by water pressure induced by high spring tides. In this case it can be concluded that oyster breeding is only indirectly related to the lunar cycle, through the gravitational pull of the Moon at spring tides.

A question remains, however, as to the biological advantages for the oysters of releasing their larvae at a particular phase of the monthly tidal, that is lunar, cycle. The process of natural selection is such that if the timing of oyster swarming is so precise, as seems to be the case, this must have occurred by progressive changes in the breeding population directed by adaptive advantage. Individual genotypes that work well in their setting can be assumed to have contributed progressively more to the gene pool and thus to the genetic profile of the present breeding population. A possible explanation for the release of larvae about ten days after the previous full or new Moon is that larvae are shed into the water about eight days after the previous spring tides, that is, during neap tides. Therefore, since the oyster beds are situated near the mouths of the Dutch rivers, larvae are released at times when they are least likely to be washed out to sea or high into the estuaries by the extensive tidal flows of several nautical miles per hour during spring tides. In this way the larvae, as they grow and settle on the sea bed, have a greater chance of doing so in the localities of the parental oyster beds and are less likely to be carried to unfavourable settling sites elsewhere. In other words, spawning during neap tides results in greater likelihood of the larvae settling in sites suitable for their survival, a tendency favoured by natural selection.

* * *

Irrespective of whether spawning in scallops and oysters is triggered indirectly by Moon-phased events such as increased water pressure or strong currents during particularly high tides, or whether rhythmic behaviour is in direct response to moonlight intensity, as seems to occur in some other animals, additional Moon-related factors concerning spawning have to be considered. Scallop and oyster breeding cycles are almost certainly partly based upon inbuilt periodicity which determines a readiness to spawn at a particular time when an appropriate environmental signal is received. If so, that periodicity, too, must have been established over evolutionary time against a background of direct and indirect environmental rhythmicity generated by the monthly lunar cycle.

Direct evidence that the timing of spawning in a species of shellfish occurs in relation to the twice-monthly pattern of spring tides is to be found in a periwinkle snail (*Melampus bidentatus*) living in salt marshes in North America. The periwinkle lives at high levels which are covered by sea water only during fortnightly high spring tides. Accordingly, since its newly hatched larvae are free-swimming, the process of hatching can only take place on a matching fortnightly basis. Hatching of the eggs is therefore synchronized with spring tides, that is, just after the times of new and full Moon. Most importantly, however, it has been demonstrated that this pattern of hatching persists as a fortnightly rhythm in periwinkles maintained in constant conditions in the laboratory.[53] Thus true adaptation to the springs–neaps cycle of tides has occurred in this species of snail, indicating indirect adaptation to the monthly cycle of the Moon.

Several other species of marine snails around the world have been observed to release their eggs during times of high tide, and many do so particularly during high waters of spring tides, again indicative of twice-monthly spawning. The common edible periwinkle of European shores is an example illustrating such behaviour. It has a lunar-tidal rhythm of spawn release, with increased output during spring tides. In this case there are evolutionary benefits of spring tide spawning because, unlike oysters that live in confined beds near estuary mouths, these creatures live on open sea coasts and find it beneficial to be widely dispersed in the currents of spring tides.

Certainly, then, modern studies justify Aristotle's view that the quality of shellfish, as determined by their reproductive condition, varies according to the lunar cycle, whether that variation in quality is directly or indirectly dependent upon the Moon itself. We also see among shellfish, as will be apparent later in other creatures, that Moon-related biological rhythms can be expressed when an animal is deprived of contact with lunar events. This means that we are not only seeing responses to lunar cycles, direct or indirect, but have evidence of the inherited capability of animals to express Moon-related behaviour when kept isolated from the environment. In other words, evidence is already forthcoming that some animals may possess internal biological clocks that cycle with the periodicity of the lunar month.

* * *

But Aristotle was not the only early observer to recognize that there are links between the availability and quality of seafood and the Moon. Many human populations living in coastal localities have been equally observant in the past—for example, concerning the annual spawning dances of the palolo worm, in waters around the Samoan Islands and in the Caribbean Sea. The notion of dancing worms might seem odd to those most familiar with earthworms exposed by a garden spade, but the folklore of some peoples living by the sea is steeped in such phenomena concerning marine worms, whose 'dancing' and availability as food is often coincident with a particular phase of the lunar cycle. Many species of marine worms exhibit cyclical patterns of reproduction that occur at particular times of the lunar month, some performing nuptial dances when males and females swim to the surface to release eggs and sperm into the sea. Fireworms on the eastern coasts of North America and ragworms on the coasts of Europe have been observed to perform such spawning rituals, which are particularly evident in the former. Normally found under rocks in coastal seas, fireworms are especially impressive at such times since they light up with bioluminescence as they reach the sea surface. In their ritual swimming behaviour males rotate their bodies around receptive females, generating small circles of light that at night are visible to the naked eye. During the dances, eggs and sperm are released simultaneously into the sea, a phenomenon that increases the chances of fertilization and successful survival of the species. In both fireworms and ragworms the whole ritual of the nuptial dances occurs just after sunset on nights immediately following the times of new Moon, coinciding with a set of spring tides.

Even more striking occurrences of Moon-related spawning are seen in palolo worms (*Eunice viridis*) around the Samoan Islands in the south-west Pacific Ocean and in a related species of marine worm around the coasts of the West Indies. In each case the worms spawn around the time of the third quarter of the Moon, the Samoan palolo in October/November and the Atlantic palolo in July. During much of the year the worms are rarely seen, since they spend most of their lives hidden in crevices in coralline rocks that fringe tropical islands, down to depths of a few metres below the sea surface. When they are seen, however, their appearance is spectacular. During the spawning season the hind parts of their bodies, containing eggs or sperm, increase dramatically in size and break off as swimming tails. In Samoa these occur in such vast numbers in surface waters near the coast that they are 'fished' by local people who regard this gift of food from the sea as a great delicacy. Moreover, local fishermen are able to predict the spawning days each year with considerable accuracy, according to a particular day and phase of the Moon on their palolo calendar. Rarely are the islanders disappointed in readying their canoes and nets in anticipation of a big annual feast. The calendar was established in folklore in early historical times, but Samoan palolo spawning is such a spectacular event that it has been validated in more recent times from the diaries of early European missionaries and medical practitioners, and subsequently by resident and visiting biologists in Samoa.

From early well-documented records in the mid-nineteenth century until the present day, Samoan 'palolo days' have been found to occur consistently on or within a day or two of third quarters of the Moon (Fig. 8) during a period of about six weeks in the months of October and November.[54, 27]

Remarkably the spawning dates follow the so-called 'metonic cycle' of nineteen-year periodicity, which arises because lunar months are shorter than calendar months. By this means, in any one year a particular phase of the Moon occurs ten or eleven days earlier than in the previous year and any one phase of the Moon, say a third quarter Moon, does not occur on the same calendar date until nineteen years later. In some years there is only one such third quarter during the six-week 'spawning window', but in other years two lunar third quarters occur, raising the possibility of two episodes of palolo spawning in a particular year. Here is an important case where folklore

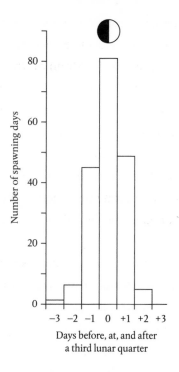

Fig. 8. Spawning dates of palolo worms, timed to the third quarter of the Moon in October/November each year.

concerning the synchronization of natural events with the lunar cycle has been validated by scientific observations. There are now over one hundred years of well-documented records that permit predictions of Samoan palolo spawning times to be made with some degree of accuracy for the benefit of local communities. For example, the palolo calendar, built up partly from Samoan folklore and enhanced by the analysis of scientific observations, predicted well in advance that the day of the third quarter of the Moon on 30 October 1980 was to be a 'palolo day'. In the event, fisheries officers of the Samoan Government

81

found that the worms did suddenly appear in vast numbers for their single spawning event of the year on the predicted day, recording in their Fisheries Report for 1980 that it was a day of 'heavy spawning strength'.[27] Moreover, the accuracy of predictions by the palolo calendar, before and after 1980, has consistently been validated by records of fisheries officers. There are only occasional years when forecasts are wrong, attributable to failure of the worms to spawn, probably associated with adverse weather conditions at some time during the physiological build up to the expected time of spawning.

As with Aristotle's urchins, it seems reasonable to conclude that synchronization of the spawning activities of palolo worms is of great advantage to survival of the species by ensuring that eggs and sperm are produced in the same place at the same time. The question remains, however, as to the advantage for survival of the species by spawning synchronously at the time of the third lunar quarter and not at some other phase of the lunar monthly cycle. The same question has been posed concerning the spawning times of sea urchins and the swarming times of Dutch oysters, and for palolo the answer may well be similar. The release of oyster larvae during the third lunar quarter, that is, during neap tides, reduces the risk of wide dispersal away from oyster beds near the mouths of estuaries, and in a similar manner palolo larvae released during neap tides have a reduced risk of being swept away from their specialized habitats around coral reefs. Indeed, some reef corals themselves also synchronize their spawning during the third quarter of the Moon. Corals off western Australia and others on the Great Barrier Reef off eastern Australia spawn at such times. Even though the former spawn during autumn and the latter during spring or early summer, in each locality the release of coral eggs and sperm occurs soon after sunset, usually about a week after full Moon. The upwards drifting 'snowstorm' of coral gametes, fertilized eggs, and larvae is a spectacular event for diving tourists and a bountiful feast for a host of marine animals that are attracted to it in a feeding frenzy.

Some species in the sea depend upon wide dispersal and colonization of new habitats for survival, and perhaps release their larvae during the strong tides and currents of spring tides accordingly. In contrast, others living as part of or in association with local specialized habitats such as coral reefs would find it disadvantageous if their young stages were distributed far from the habitat where they thrive most successfully. Release of palolo larvae during neap tides, like those of corals themselves, would therefore be less likely to result in larvae being swept away from fringing reefs, into deep ocean localities where suitable settlement sites would not be available. Such an explanation for quarter-Moon timing has yet to be verified by demonstrating that it enhances the survival chances of the species concerned, but the circumstantial evidence for palolo and other animals is supportive. In evolutionary terms, Samoan palolo worms can be considered to have weighed up the costs and benefits of wide dispersal of their larvae, 'resolving' that staying put during neap tides of a lunar quarter is of greater advantage for survival than wide dispersal during spring tides of full or new Moon.

However, the story has to be more complicated than that, since palolo spawning occurs only during neap tides of a third quarter of the Moon and not during those coincident with a first lunar quarter. Pending further study of this phenomenon, a constructive suggestion that can be put forward is that palolo worms as they prepare to spawn in some way respond directly to moonlight. It can be hypothesized that as the 'spawning window' opens each year, the worms perceive the high intensity of moonlight during full Moon nights as a trigger that sets in motion the final stages of maturation of eggs and sperm which are then released some days later during the following third quarter of the Moon.

Whatever the biological advantages of Moon-related spawning activities in marine worms, phased indirectly by tides or directly by moonlight, observations of such phenomena have been sufficiently convincing to encourage some biologists to investigate such findings in carefully designed experiments. For example, studies have shown that the dawn and dusk foraging behaviour of ragworms on British shores is cued by extremely low levels of light. They respond to a period before dawn when light is imperceptible to the human eye, at levels well below that of moonlight in shallow sea water.[55] More directly, the spawning behaviour of Atlantic palolo worms (*Eunice schemacephala*) has been shown to vary in relation to changes in moonlight intensity. At the time when Atlantic palolo were expected to spawn, exposure to light intensity equivalent to that at full Moon induced light-avoidance behaviour in the worms. They retreated into rock crevices and did not spawn when moonlight was very bright. The worms simply ignored light intensities equivalent to those of new Moon but were attracted out of their burrows to release their swimming tails when exposed to light intensities equivalent to those of lunar quarters.

The responsiveness of marine animals to moonlight is an intriguing and challenging field of study and probably the first person to approach the subject in a comprehensive manner was a German biologist, Carl Hauenschild.[56] He chose to work at the Stazione Zoologica, an attractive laboratory for research visitors, situated as it is on the shores of the Bay of Naples, in Italy, where moonlight is a more predictable phenomenon than at laboratories, say, on the shores of northern Germany. Hauenschild began by confirming earlier anecdotal accounts that a species of bristle worm (*Platynereis dumerilii*) living in the Mediterranean Sea spawned most abundantly at times of new Moon, performing nuptial dances as they did so. He established that at sexual maturity the worms transform into breeding mode, the stage at which they swim up from the sea bed, swarming at the sea surface before they spawn and die. Then, as a basis from which to carry out experiments to understand more fully the timing of swarming activity, his first task was to culture the worms in tanks in the laboratory, exposing them to the normal day/night cycle and to monthly changes in moon-light at night, exactly as if they were in the sea. By this means it was possible to establish that they bred in the laboratory in an identical manner to their counterparts in the sea, performing ritual nuptial dances on time during nights of the new Moon.

Then, using the question and answer, or hypothesis-testing approach of science, it was necessary to test how a laboratory culture of the worms would behave if they were continuously illuminated throughout the day and night, and were therefore oblivious to changing moonlight. Under these conditions it was found that the worms spawned at irregular intervals throughout the lunar month. They showed no preference to swarm and release their eggs and sperm at a particular phase of the Moon. Worms kept in an artificial light/dark cycle with no reference to the Moon were similarly irregular in their spawn-ing, but the next question was whether such a culture could be induced to re-acquire lunar periodicity by exposing it to moonlight during the dark phases of the laboratory light/dark cycle. To answer this question such aquarium cultures of worms were, on a few successive 'nights', exposed to a light intensity equivalent to that of full moonlight. The experiment proved a success because subsequently the cultures were again found to spawn rhythmically, with greatest spawning intensity timed to an interval starting about two weeks after the worms were last exposed to artificial moonlight. Moreover, in repeat experiments, it was shown that peak spawning began at the same time after the last night of exposure to artificial moonlight, irrespective of the number of artificial 'moonlit nights' to which the worms were exposed. Since the phase of the new Moon begins about two weeks after the full Moon, it is tempting to suggest that, in nature, the light of the full Moon in some way sets in motion physiological processes within the worms which lead to their nuptial dances and spawning around the times of new Moon, fourteen or fifteen days later.

Intriguingly in these experiments by Hauenschild, exposure to a few successive 'nights' of artificial moonlight not only triggered maximum spawning by the worms about two weeks after the event, but the experience also set in motion episodes of spawning at long intervals thereafter. After the initially induced spawning event, several further maxima of spawning occurred at intervals of twenty-eight or twenty-nine days thereafter, a periodicity which closely approximates to the lunar monthly cycle. Sceptics of the influence of the Moon on living animals must here take note. Not only did these bristle worms appear to show a clear response to changing levels of moonlight, but that response seemed to set in motion within the worms an inbuilt mechanism of Moon-related physiology which in nature controls the timing of the spawning dances of the worms at new Moon. A conclusion to be drawn from such a finding is that *Platynereis* has been exposed to cycling of the Moon for long periods of evolutionary time to the extent that it has acquired internal biological clockwork, at the level of individual cells, that approximately matches the periodicity of the lunar month. Indeed, a modern discovery supports that conclusion, finding that circalunar biological clockwork in these bristle worms has a molecular, that is, genetic basis,[31] as we will see later in the book.

The biological advantage of cellular clockwork of approximately lunar monthly periodicity that can be repeatedly synchronized by exposure to moonlight at full Moon is that it is able to continue running, approximately in time, on occasions when the worms are deprived of moonlight signals during overcast nights. Similar Moon-related rhythms of reproduction which can be manipulated by exposure to artificial moonlight have been well documented in other animals, and indeed plants, from seaweeds to crabs and insects.[27] In some cases breeding maxima occur twice each lunar month, rather than once, but the principle of control by inherited molecular clockwork synchronized by changes in moonlight intensity seems likely to be widespread.

In this chapter we have seen that some scientifically proven, modern examples of Moon-related rhythms in animals have their origins in folklore, validating a comment by Arthur Winfree in his 1987 seminal book *The Timing of Biological Clocks,* that 'important discoveries are sometimes made intuitively or by diffuse folk wisdom before formal proof can be mustered'.[57]

In the next chapter we will explore more instances where seemingly mythical stories of Moon-related biological events indeed have scientific validity.

5

STRANGERS ON THE SHORE

Throughout the centuries, in many parts of the world, human populations living by the sea must have been aware that, at certain times, large numbers of normally sea-dwelling animals suddenly emerged from the waves, to be apparently stranded between the tides.

The strandings would in some cases be gifts of food from the sea, often as random occurrences following storms. In other cases, however, the apparent stranding events would, over time, have been recognized as occurring in a pattern, at certain times of the year and at certain times of day. Eventually it would have been apparent that many of the strandings were in some way timed to phases of the Moon. Today we have substantial documented evidence of such a connection.

Nowadays it is well known that several kinds of oceanic animals come ashore in this way, usually to breed. Among these, the variety of sea turtles to be found coming ashore in a number of localities around the world are probably the best known. As a group, turtles have lived on Earth for over 150 million years and, from evidence of the fossil record, they first evolved when reptiles in general, including the dinosaurs, were particularly abundant during the Jurassic period of the Earth's history. Now they occur almost anywhere in warmer seas, typically coming ashore to lay eggs on tropical and sub-tropical sandy beaches around the world. Though they are air-breathers, over their evolutionary history some have acquired the habit of feeding in the open ocean on unlikely prey such as drifting jellyfish, while others graze on coastal sea-grasses before returning to deposit their eggs on the beaches where they were born.

Some green turtles (*Chelonia midas*) navigate very precisely over distances of about 2,000 kilometres from sea-grass beds on the coasts of Brazil to breed on beaches of the small mid-Atlantic island of Ascension.[58] As the migrating male and female turtles approach their target beach, mating occurs in the sea, before the males reverse their swimming direction and move out to sea again. Meantime the females congregate in the shallows along the coastline, preparatory to their synchronized emergence on to the beach to lay their eggs. Ashore they move laboriously to the top of the beach where they excavate a depression in the sand into which they deposit their eggs before covering them and returning seawards again. They are usually ashore for only an hour or two, mostly at night and often at high tide, laying eggs in dry sand, probably as a legacy of their ancestral origins from terrestrial reptiles.

Strikingly, however, in many localities around the world, the various kinds of turtle all seem to time their emergence from the sea fairly precisely around the times of full Moon and perhaps the new Moon too. Typically, they emerge during darkness when the risk of attack by bird predators is least, and during full or new Moon high spring tides when the distance to haul themselves to the nest sites is the smallest. But the association of turtle egg-laying with the Moon does not stop there. It is known that in some localities females on their return journey to the sea after depositing their eggs orientate themselves according to the light of the Moon reflected from the surface of the ocean to which they are heading. Hatchling turtles, as they emerge from the eggs to make their way seawards at a later date, also orientate themselves in the same way. There is circumstantial evidence that turtles utilize a nocturnal light source for navigation in this way, since up-shore and down-shore migrations have been seen to be disrupted on turtle nesting beaches where artificial lighting is present in buildings along the coastline. Some resident coastal migrant species of other animals, such as sandhoppers, have also been reported to be similarly adversely affected in their homing behaviour on beaches in California, sometimes by so-called 'morality lighting' installed to deter amorous human behaviour on beaches at night.

Relationships between the phases of the Moon and the timing of turtle breeding are no doubt biological rituals established over long periods of evolutionary time. It seems likely that they were established before the appearance of birds, let alone man, in the evolutionary history of the planet. Yet birds and humans are now serious predators of turtles, their eggs and their hatchlings, to the extent that many species of turtle around the world are under threat of extinction. In the early millennia of the existence of turtles, before the appearance of birds and humans, evolution of this elaborate breeding cycle must have been particularly advantageous, albeit necessary for a group of animals with terrestrial ancestors, but the arrival of man has clearly been disadvantageous. Humans appear to have been aware of the association between turtle breeding and the Moon. It is captured in phrases such as 'The Turtle Moon'. Early native American culture recognized the association by characterizing the thirteen plates of the shell of some turtles as the thirteen lunar months of a year. Improved understanding of the predictability of Moon-related behaviour of turtles no doubt led to their over-exploitation as human food. Hopefully that understanding can in future be used to help establish management rationales for their conservation.

* * *

Another intermittent visitor on the shores of the eastern USA, the Gulf of Mexico, and various countries in Asia, including Japan, Malaysia, Indonesia, and the Philippines, is the so-called horseshoe crab, *Limulus polyphemus*. This is not a crab at all but an arachnid, more closely related to spiders and scorpions, which has remained relatively unchanged for an even longer period of geological time than that survived by turtles and their ancestors. Fossils virtually identical to modern horseshoe crabs date back to the Triassic period of the Earth's history, around

220 million years ago. Its young stages are referred to as 'trilobite larvae' in deference to their shield-like carapace that gives them a superficial similarity to fossil trilobites. Not surprisingly horseshoe crabs are sometimes referred to as 'living fossils', an unfortunate contradictory term for one of the great surviving groups of early evolution. They have lived through several cataclysmic episodes of extinction of organisms during the Earth's history, not least that of about 65 million years ago which contributed to the extinction of the dinosaurs.

Off the east coast of North America, horseshoe crabs are normally resident at some depth, where they scavenge on the sea bed for worms, molluscs, and fragments of algae as food.[59] Like turtles they also come ashore on sandy beaches to breed, though in this case both males and females emerge as the strangers on the shore. When this happens, clusters of the animals are to be found, usually with a large female partly buried as she lays her eggs, which are fertilized by several smaller males clustering around and on top of her.

A single female may lay several thousand eggs, which, after fertilization, are then buried by her until they eventually hatch as 'trilobite larvae', measuring about one centimetre in width, which commonly live for a time on sand-flats between tide-marks. In such habitats, the larvae grow to a width of about four centimetres in a year before eventually moving offshore, where they join the adult population. There they mature slowly and return to spawn after about ten years, living for a total of about twenty years after hatching.

Like turtles, the arrival on the beach of copulating horseshoe crabs can usually be predicted. On the east coast of the USA it tends to occur fortnightly in spring and summer, mostly around the times of new and full Moon high spring tides. As in turtles, the pattern of emergence probably evolved millions of years ago, long before the appearance of birds in the evolutionary record, yet birds are now major consumers of the newly laid eggs of the 'crabs'. In some localities around the world, humans too collect horseshoe crabs for food, notwithstanding the fact that they contain little suitable flesh for eating. Here then, as in turtles, Moon-phased co-ordination of accumulations of the 'crabs' leads to predictability of their behaviour which exposes them to modern threats. As with turtles, human threats now potentially affect the survival of horseshoe crabs, another group of animals that hitherto have thrived through aeons of evolutionary time.

* * *

Yet another, classical and spectacular example of the phenomenon of Moon-related spawning behaviour that brings a marine species ashore concerns the grunion. This is a small sardine-like fish, *Leuresthes tenuis*, which emerges from the sea to spawn on some sandy beaches on the coasts of southern California and western Mexico.[60] Grunion normally live in the open sea but show themselves in an annual spawning frenzy when they come ashore for egg-laying and fertilization, usually in April and May. At the time of spawning, male and female fish rush inshore on to sandy beaches, riding incoming waves at night, to shed their eggs and sperm at the highest levels of the beach. At high tide level, partly out of water and exposed to the air, the fish dig pits in the sand, into which they deposit eggs, followed quickly by sperm for fertilization, before male and female fish rush seawards again. Again the spawning frenzies are timed to occur very precisely during a period of one to three days just after the times of full and new Moon at spring tides. Indeed, the timing of spawning is so well correlated with the timing of spring tides, and hence with particular phases of the Moon, that predictions of grunion spawning are regularly published in local newspapers during the expected breeding season.

The Moon-related behaviour of grunion also extends to the newly hatched fish, which have to make their way seawards. After being laid and fertilized, the eggs hatch whilst buried in sand near the high-water mark, and they develop rapidly as juvenile fish during the next fourteen or fifteen days until the following spring tides, withstanding some drying out when they are not covered by intervening neap tides. By the time of the next session of spring tides the juvenile fish are ready to swim and are carried seawards, making the reverse journey to that made by the adults at the time of spawning. Individual fish are known to spawn during successive spring tides, whether timed to new or full Moon, so it seems that spawning is timed not directly to the Moon but indirectly so through the effects of the springs/neaps tidal cycle. It is possible that such precision of reproduction is determined by a physiological rhythm of annual reproductive periodicity which interacts with semilunar rhythms of gonad maturation at the time of spawning, which are themselves cued precisely by a tidal factor such as high water pressure during a previous spring tide. However such precision is achieved in the grunion, its mode of reproduction is highly unusual among fishes. Most marine fish spawn in the sea, where their eggs are at risk of being eaten by any number or predators, including other fish. It has to be assumed, therefore, that throughout the course of evolution grunion have come to spawn at the very fringes of the ocean as an adaptation to avoid the attentions of ocean-living predators. In a sense they have balanced the benefits of avoiding predation of their eggs against the risks of stranding and predation of adults and young fish as they flounder about between tidemarks during or after spawning. Again, though, they probably evolved this Moon-related breeding strategy before the appearance on the evolutionary scene of predatory birds!

* * *

Not all animal strangers that appear on the shore emerge intermittently from the sea; some species migrate there from inland habitats to breed at the edge of the sea. In the gee-whizz presentations of some natural history programmes on television it is not uncommon to see images of land crabs heading towards the sea, startling people in their homes and gardens, and running the risk of being crushed by passing vehicles on coastal highways (Fig. 9). What we are not always told is that

Fig. 9a. Red crabs on Christmas Island undertaking Moon-phased migrations from their inland burrows to shed their larvae in the sea, crossing highways and rail tracks as they do so.

such events often coincide with a particular phase of the Moon.

Such terrestrial crabs occur in tropical and sub-tropical localities around the world and in some places are known to live up to several kilometres inland from the sea. Yet their life-cycle can only be completed when female crabs carrying eggs return to the sea to release their offspring. The young are swimming larvae that must undergo their early development in the watery medium of the sea, towards which mature females are lured when their eggs are ready to hatch. On land,

Fig. 9b. Continued.

the adult crabs, which are not well adapted to respire and survive in air, usually burrow to depths where they reach the water table, enabling them to keep moist. Accordingly they normally avoid the risk of desiccation during the day by emerging to forage at dusk and dawn, when, in some parts of Central America, they are collected for food. After mating, the females are able to carry eggs for up to several weeks, and when the eggs are ready to hatch the crabs migrate, often over a period of one or two days, until they reach the sea, where the eggs hatch and the larvae are released. In the open sea the larvae undergo development for several weeks before they return ashore, moulting to the crab stage as they do so. Crucially, in some regions of the world the timing of migration of egg-bearing females is often within a day or two of the full Moon, or of the full and new Moon, depending upon the locality. In each case the newly swimming larvae are released at the water's edge during high spring tides. At those times the distances travelled by females over the open beach to reach the edge of the sea is less than would be the case if they migrated during smaller neap tides. Moreover, once released, the larvae are at less risk of stranding between tidemarks by being carried quickly offshore by the strong currents of spring tides.

There are, however, exceptions to this pattern of release. In some regions of the world, particularly on oceanic islands, it could be disadvantageous for larvae to be distributed widely in the ocean on the strong currents of spring tides. There they may be carried far offshore to open waters where the chances of finding a suitable landmass on which to settle would be virtually zero. On such isolated islands, reflecting the ability of animals to adapt to local conditions, the timing of release of larvae by the parent crabs tends not to be associated with the new and full Moon, but with times of the quarter Moons. At those times, of neap tides, tidal currents are at their weakest, increasing the chances of larvae remaining near the island from which they were first released into the sea.

One example of a land crab showing this type of behaviour on a remote oceanic island is *Johngarthia lagostoma*, which is found on the mid-Atlantic island of Ascension. The annual breeding migration of this crab, from inland to coastal beaches, occurs from January to May, peaking in March in most years, with increased numbers appearing on the shore during the first quarter of the Moon and even larger numbers during the third quarter.[61] Another typical example of such evolutionary novelties is the Christmas Island red crab, *Gecarcoidea natalis*. This is a land crab that is found only on Christmas Island and the Cocos Islands, two remote but relatively close small land masses situated south of the Indonesian island of Java and north-west of Australia in the Indian Ocean. On Christmas Island alone millions of red crabs are found living in burrows inland, emerging at dawn to feed on vegetation and, remarkably at particular times of year, migrating *en masse* to the sea to mate, lay their eggs, and release their larvae into the sea.[62] The crabs walk for several days, covering distances of several kilometres from their normal habitat in the high rainforest plateau, eventually burrowing and mating just short of the sea coast, before females continue seawards to release their larvae at the water's edge. The highly synchronized migration of millions of breeding crabs occurs in the monsoon season, during late November or early December, timed so that larvae are released into the sea during a lunar quarter when tides are small and tidal currents around the island are weak. Accordingly the crabs are endemic, that is unique, to Christmas Island and the Cocos Islands, and do not occur on Indonesian shores to the north. This isolationism increases the risk of inbreeding, of course, but that is a characteristic of animal populations living on isolated islands. Such isolated populations tend to evolve as separate species, as Charles Darwin noted in his studies on the Galapagos Islands, and likewise Alfred Wallace, in his studies in the Malay Archipelago, as they formulated the concept of evolution by natural selection.

In mainland Japan, seaward migration is undertaken by land crabs that live no more than 200 metres or so from the sea, but in other localities such crabs may live much further inland where the sea coast might seem relatively inaccessible. In such cases some terrestrial crabs avoid long overland migrations by making their way to streams and rivers to release their larvae, which are then carried to the sea by river discharge. Even so they observe precise timing in relation to the phase of the Moon; they migrate to release their larvae during high spring tides, around full and new Moon, again suggesting that they benefit from having to walk as short a distance as possible to spawn.[63] The crabs exploit the fact that water flow at the river mouth backs up with the incoming spring tide, flooding far up the river bank as it does so. This supports the idea that the crabs are truly timing the pattern of their behaviour in some way to the light of the Moon and not to some distant signal, say, from the crashing of waves on a sea shore. Indeed, this has been confirmed for the land crab, *Sesarma haematocheir*, by M. Saigusa in Japan.[64] In laboratory experiments it was shown that a fortnightly rhythm of spawning could be set in motion in a randomly spawning group of these crabs by exposing them to simulated moonlight at night. In other words, the crabs seem to have in-built biological clockwork, the timing of which can be set by the light of the Moon, to enable them to make their rare appearances at the edge of the sea in anticipation of the times of fortnightly high spring tides during their breeding season.

But it is not only land crabs that show Moon-related behaviour that brings them to the edge of the sea at particular times of the year. Fiddler crabs, which are so common on tropical and sub-tropical beaches, often extend their ranges into the freshened water of estuaries, marshes, and mangrove swamps, from where they too need to migrate to the edge of the sea when their eggs are ready to hatch. Again, in some localities, they can be seen to do so in large numbers during the highest spring tides around the times of full and new Moon. Moreover, in view of the fact that in some localities fiddler crabs may live in moist situations that are not normally reached by tides, it seems likely that they, like land crabs, are cued by the light of the Moon in their fortnightly migrations to the edge of the sea to release their larvae.

* * *

Yet more strangers that appear intermittently on the shore in relation to the lunar cycle are swarms of midges. Midges are ubiquitous insects, renowned for their swarming behaviour and the nuisance they cause to hikers and campers in some parts of the world. Less well known, perhaps, are the midges one occasionally sees in vast swarms when exploring rocky sea shores of some European and other coastlines. It is highly unusual to associate insects with the marine environment, except perhaps among rotting seaweed washed up above the level of high tide. Early in evolution the ancestors of this group of invertebrate animals acquired the ability to breathe air as they colonized dry land habitats, and only a few have reverted to breathing in water. While some have colonized fresh waters, few have reverted to breathing in sea water, particularly as adults. Yet there is an apparent exception, where, in some coastal localities, swarms of adult midges can sometimes be seen surging into the air from amongst inter-tidal rocks revealed by the falling tide. The timing of swarming by such coastal midges is Moon-related, occurring with high predictability, approximately every two weeks. During the breeding season swarms of midges are to be seen low down on a shore, near the water's edge when the tide is out during the fortnightly spring tides, at times just after new and full Moon.

Those marine midges that have successfully re-colonized the coastal zone, to appear intermittently as strangers on the open shore, have done so in an unusual way. They spend most of their lives as eggs, larvae, and pupae hidden in crevices and among living seaweeds, during which parts of their life-cycle, when exposed to the air at low tide or when immersed in sea water at high tide, they survive by absorbing oxygen across their body surface. The adults, when they emerge from the pupae, must breathe air, which on the face of it is not a problem for the males, which are able to fly, but it does present a problem for the females, which are flightless. Accordingly, since males and females must remain in close proximity for mating, it turns out that both males and females are able to survive for only a few hours during low tide before they are drowned by the next rising tide. During this very short interval of time they must achieve mating and egg-laying before they inevitably die, and it is this brief but intense mating ritual that is observed as swarming behaviour at low tide. The swarms of flying midges are, of course, all males seeking to mate with the non-flying females, which are often to be seen helpless on the water surface of rock pools at low levels of the shore. After mating, egg-laying then takes place on seaweeds and similar surfaces before males and females meet their fate with the flooding tide. Then the life-cycle begins again, with eggs, larvae, and pupae surviving successive periods of immersion by flooding tides until maturity, when the mating ritual begins again.

The significance of Moon-phased timing of swarming becomes apparent when it is appreciated that the zone of occurrence of the midge eggs and larvae is very low down on the shore, at levels that are exposed to the air for only a few days, twice each lunar month, during low spring tides. During neap tides the lowest levels of a shore are never exposed to the air throughout a tidal cycle, ensuring that animals and plants at those levels, including the midge larvae, remain continuously submerged at those particular times of the month. Emergence of the flying males can therefore occur only during low spring tides, which, as we have seen, in any one month occur just after the days of new and full Moon.

So far, as with many reports of Moon-related phenomena, the story is anecdotal, albeit in this case with an explanation that is consistent with sound knowledge of the biology of shore-living midges and the physical basis of tides, in relation to which midge lifestyle can be considered to have evolved. But questions arise for the curious biologist. Exposure to the air during low spring tides is presumably the direct trigger for males to hatch and fly in search of receptive females, but what determines that male and female midges are in the appropriate reproductive state to be able to respond at the relevant time when the tide is fully out? Is the periodicity of their reproductive cycles determined by previous experience of the monthly cycle of tides, or are the midges in some way responding to the monthly cycle of the Moon? In other words, are their reproductive cycles related indirectly or directly to the Moon?

Low spring tides, as we have seen, are caused by the gravitational effects of the Moon and the Sun when they are in alignment with the Earth, and their fortnightly repetition might seem to present the most logical basis from which to explain the pattern of mating swarms of coastal midges. Certainly that explanation might be accepted without challenge by anyone sceptical of the possibility of the direct involvement of moonlight itself in helping to phase the swarming ritual. However, science progresses by asking challenging questions and these came to be asked when it became possible to culture shore-living midges (*Clunio marinus*) through their complete life-cycle in the laboratory. This provided Dietrich Neumann, a German researcher at the University of Cologne, with the opportunity to investigate the mating behaviour of midges in controlled conditions and to assess how that behaviour varied in response to a range of simulated environmental conditions.[65]

First, midges from the coasts of Spain were cultured in the laboratory in a standard day/night cycle of twelve hours light and twelve hours dark, away from the influence of tides. Under these conditions male midges were found to emerge and fly in search of females with which to mate, but they did so with random timing. They showed no fortnightly pattern of swarming that was equivalent to the pattern shown by midge populations on their native beach. However, it was found that the fortnightly periodicity of the mating ritual exhibited by midges in nature could be replicated in laboratory midges by a simple procedure. All that was required was to expose a culture of

Clunio to artificial moonlight during four nights of the 24-hour light/dark cycle to which they were exposed when being cultured in the laboratory. If the four nights of simulated moonlight were repeated again after 29 and 58 days, exactly as if the midges were exposed to the monthly cycle of full Moon, the natural fortnightly rhythm of mating ensued. Moreover, the same twice-monthly pattern of mating flights continued, even after the artificial cycle of moonlit nights was discontinued. Quite clearly these experiments seemed to provide evidence to support the notion that in nature moonlight had a direct and persistent effect on the timing of the fortnightly mating rituals of the midges in question. It seemed, too, that *Clunio* might possess internal biological clockwork of fortnightly periodicity that was cued by the Moon.

But the story is even more impressive than it might at first appear, since a persistent fortnightly mating rhythm could also be induced by exposing a culture of midges to artificial moonlight for just three nights in succession (Fig. 10) or even for one night only.

In the experiments involving exposure to moonlight for one night only it is clear that no information was conveyed to the insects concerning the periodicity of a moonlight cycle, yet a persistent fortnightly rhythm of mating was set in train. That discovery led to the more definite conclusion that each midge possesses some form of biological clock of fortnightly periodicity. During the breeding season, such a biological clock would ensure that the insects as a population are in an

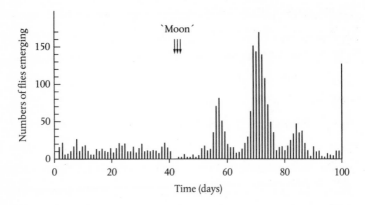

Fig. 10. Midges (*Clunio marinus*) cultured in the laboratory in a 24 hour light–dark cycle hatch randomly until exposure to artificial moonlight during three nights induces a semilunar (14-day) hatching rhythm.

appropriate physiological state to breed when exposed to the air during low waters of spring tides, ensuring that these strangers on the shore were there under their own volition. The fortnightly biological clocks of individual midges would be out of phase with each other when laboratory cultures are maintained in conditions of light and dark. However, the clocks of all the insects would be synchronized by brief exposure to artificial moonlight during a dark 'night' phase in the laboratory, to set in motion the normal fortnightly pattern of synchronized mating behaviour.

Here, then, in coastal-living midges, we have more evidence, derived from convincing experiments, that moonlight has an effect on some living organisms and that the lunar effect is direct, through changes in moonlight intensity throughout the lunar monthly cycle.

But the question remains as to how universal the phenomenon is, even among coastal midges. The midges studied in
Neumann's experiments discussed so far were from the coasts
of Spain, where moonlight might be expected to be more
predictable than, say, on the coasts of northern Europe, where
populations of *Clunio marinus* are also known to occur. On
northern European coasts midges can indeed be seen swarming
in mating flights during fortnightly low spring tides. However,
unlike their Spanish cousins, they do not respond to artificial
moonlight when cultured in the laboratory. In those localities
the midges show only indirect responses to the Moon by timing
their fortnightly clockwork to the rise and fall of tides. Experiments by Neumann that confirmed this interpretation of events
were carried out on populations of midges from Helgoland in
northern Germany and from Normandy in northern France.
Midges in both localities performed their mating rituals during
the hours of low spring tides, but the timing of low spring tides
in the two localities is different. Of course the Moon would be in
the same phase at each locality at the same time of day, but the
timing during the day of low spring tides differs in each locality
because the pulse of tide takes time to travel eastwards and
northwards along the English Channel and into the North Sea.
Accordingly there is an interval of two hours between the times
of low (and high) tide in Normandy and Helgoland, exactly the
time interval between midge mating rituals in the two localities.
This difference in the timing of mating flights in relation to tides
in the two localities can also be shown quite clearly in cultures of

midges kept in the laboratory. In these two localities the fortnightly clocks of the midges appear, therefore, not to be set by moonlight, but probably by the time of the day when they are exposed to wave action during low spring tides. This was confirmed by experiments in which the cultures of midges from the two localities were subjected to stirring at different times of day, simulating wave action that they would experience at low tide.[66] In this way, the timings of mating flights in the laboratory that mimic those in nature were achieved only when artificial tidal stirring was applied at times of day consistent with the timing of low spring tides in the two localities concerned.

So, unusual appearances of animals on the sea shore, from midges to horseshoe crabs, true crabs, fish, and turtles, can be seen to occur with great predictability, commonly related to the cycling of the Moon. However, studies of such occurrences, particularly relating to the lowly midge, may also offer a salutary tale that widens our interpretation of the ways in which we perceive the Moon to be a driving force in our understanding of the ways in which seemingly strange animals are found intermittently, breeding between tidemarks. Is it possible that such breeding behaviour is most likely to be directly related to the Moon in low latitudes, where moonlight will be a fairly predictable feature of the environment? In contrast, in higher latitudes, where clear skies and moonlight are less predictable, is it likely that behavioural cues are more likely to come from indirect effects of the Moon through the influence of tides? The experiments with *Clunio* certainly appear to support that conclusion. But another question also now arises, as to whether animals from non-tidal localities are able to exhibit Moon-related rhythms, in the absence of indirect effects of the Moon through tidal influences. We address this question in the next chapter.

6

MOON-RELATED BIOLOGICAL RHYTHMS WITH AND WITHOUT TIDES

It will be apparent in the book so far that there are many aspects of marine animal behaviour, particularly breeding behaviour, that can be correlated with the periodicity of the Moon. It will be apparent too that some of those correlations may be causally related, experimental studies in some cases suggesting that periodic changes in the intensity of moonlight may directly serve as an environmental synchronizing agent that helps to set the timing of breeding periodicity. In other cases, however, it seems more likely that correlations between monthly changes in the intensity of moonlight and breeding cycles are indicative only of an indirect relationship between the Moon and breeding. In those cases the causal linkages are effected through intermediate environmental factors induced by Moon-driven tides. How extensive, then, are the direct effects of the Moon in controlling animal behaviour, compared with indirect effects induced by intermediate factors that vary with lunar cycles throughout the day, month, and year?

This is a key question for sceptics who might question whether there are any direct links between the Moon and marine animal behaviour. They might suggest that such apparent links arise only because of indirect lunar effects through the influence of tides. We have seen that this dilemma is apparent in several of the examples discussed so far, and there are many other examples of studies of apparent Moon-related behaviour in which the same question can be raised. For example, a particular phase of the monthly cycle appears to trigger the time of arrival of the young stages of some crustaceans as they come inshore after swimming in the plankton for some weeks after they have hatched. This is seen in the larvae of a spiny lobster, *Panulirus cygnus*, which occur hundreds of kilometres offshore from the coasts of western Australia in the south-east Indian Ocean, to which they must return to join the parent stock at the time of metamorphosis to the juvenile lobster stage. Ocean currents eventually carry the larvae eastwards towards the coast for several weeks, after which they come ashore as juveniles in greatest numbers during spring high tides associated with the times of new Moon.[67] There they join the parent stock that forms a commercial fishery resource which is of high value to the Australian economy. Whether their arrival is cued directly by some aspect of the lunar cycle or indirectly by monthly variations in tidal height is a question addressed by fisheries biologists studying the lobsters, but the question remains unresolved. However, direct response to low intensity of moonlight during new Moon seems to be at least a possibility, otherwise why do they come ashore only during new Moon spring tides and not also during spring tides during full Moon?

In fact we have looked briefly already at the problem of direct or indirect effects of the Moon, when discussing the breeding behavior of sea urchins in the context of the *syzygy inequality cycle* (SIC). You may remember that this is the cycle during which the largest tides alternate between new and full Moon high spring tides with a periodicity of fourteen months. In the sea urchin study it was found that the urchins ignored the SIC of tides but maintained their breeding periodicity in direct relationship to the Moon. However, the effects of the SIC on marine animals in general are not trivial. This became apparent in a study of the breeding times of crabs that live between tidemarks on mangrove-covered shores in East Africa. It was known that fiddler and grapsid crabs hatched and released their larvae there during spring tides with monthly periodicity, seemingly indifferently at either full Moon or new Moon. Only after careful study was it appreciated that larvae were always released at the times of the highest spring tides, which varied in their occurrence at times of new Moon or full Moon according to the SIC.[68] In all, six species were investigated: three fiddler crabs (Ocypodidae), *Uca annulipes, U. inverse,* and *U. vocans,* and three sesarmid crabs (Grapsidae), *Perisesarma guttatum, Chiromantes ortmanni,* and *Neosarmatium meinerti* from mangrove creeks in Kenya and Tanzania. Eleven populations of crabs were studied, from the two locations, and eight of these clearly synchronized the release of their larvae with the SIC. The crabs in question do not migrate to the water's edge to release their larvae but do so from the shelter of their burrows high on the shore between tidemarks, at all times waiting for the highest tides to reach them. So the crabs, unlike the sea urchins, did not ignore the SIC, switching their times of larval release between times of full and new Moon as dictated by the SIC of tides, accurately following the fourteen-monthly cycle during which highest spring tides switched between new and full Moon. Larvae were therefore always released at the times when tides reached the highest levels of the shore, optimizing their chances of dispersal by strong tidal currents into the open sea, where they undergo their early stages of development. Martin Skov, who, with other European and African colleagues, carried out the study,[68] succinctly summarized their findings with the statement that 'crabs synchronize reproduction to a 14-month lunar-tidal cycle'.

So, a partial answer has already been given to the question of whether lunar rhythms in marine animals are directly or indirectly determined by the Moon, by comparing the breeding rhythms of some sea urchins and mangrove crabs against the subtlety of the SIC of the Moon and tides. Sea urchins have evolved the capability to respond directly to the Moon, and mangrove crabs respond only indirectly to the Moon through its influence on tides. It also seems likely that other organisms in the sea have solved the problems associated with breeding periodicity in one or the other of these ways. Crabs on the Pacific coast of Panama, shore-living snails in Malaysia, and Gulf killifish in Florida all show breeding periodicities that switch between new Moon and full Moon spring tidal dominance, probably linked to the SIC.[68] But what determines the *syzygy inequality cycle*, the nature of which has only been alluded to hitherto?

The phenomenon of close alignment of the Sun, Moon, and Earth that occurs twice each synodic month at times of full and new Moon is known as *syzygy*. During this cycle, when the Moon is in alignment between the Earth and the Sun it produces new Moon spring tides, and when it is in alignment on the far side of the Earth it produces full Moon spring tides, on the basis of which it might be thought that the effect of the Moon would be to produce high spring tides of more or less equal height during springs and neaps. This would occur if the orbit of the Moon was circular and in a plane with the Earth's equator, but it is not. The Moon's orbit is elliptical and tilted such that the distance between the Moon and the Earth varies cyclically, causing differences in tidal heights at various times during lunar orbits. It is the interplay of this cycle of the Moon's proximity and the duration of the cycle of the true synodic month (*syzygy*) from one full Moon to the next that generates the fourteen-month *syzygy inequality cycle* (SIC). During the cycle the highest spring tides occur in association with full Moon for about seven months, followed by coincidence with new Moon for the next seven months (Fig. 7), and so on repeatedly in a pattern matched by crab breeding.

But the subtlety of linkages between the behaviour of crabs and the Moon extends beyond the responses of mangrove crabs to the SIC. The common green shore crab, *Carcinus maenas*, which has a wide, cosmopolitan distribution, has been found to show distinct Moon-related periodicity in its pattern of moulting during early stages of its life history.[69] When the larvae of this crab species are ready to descend to the sea bed from open sea plankton as juveniles they congregate in large numbers at the water's edge. They then come ashore during the high tide of spring tides, ensuring that they are most commonly found under stones at the highest reaches of the beach during the early stages of their development as small crabs. As newly settled young crabs, at such high levels of the shore, they tend to be fully immersed in sea water only at times of subsequent high spring tides associated with the new and full Moon. Between times, during neap tides, they may not be covered at high tide, requiring them to remain hidden in crevices or under stones where they avoid desiccation. With such dramatic changes in the conditions of their tidal environment, and their fragile nature, it is clear that the most advantageous times for them to moult and grow are during high spring tides associated with new and full Moon, when they are fully immersed in sea water. But they moult and grow at such times not simply because they are covered during the highest tides. Strikingly, it has been found, by observing newly settled crabs that have been transferred to laboratory aquaria, that even these moulted with greatest frequency in a fortnightly pattern, with large numbers growing by shedding their shells quite spontaneously in a distinctly rhythmic pattern at times just after full and new Moon.[69] These timings, clearly based on the possession of internal biological clocks of semilunar periodicity, are coincident with times of the highest spring tides that the crabs would have experienced had they remained on their native beach, when support by immersion would have been beneficial for them to moult (Fig. 11).

In the brief time between shedding an old shell and hardening the underlying new carapace, so-called 'soft crabs' are particularly fragile and it is advantageous for them, when in

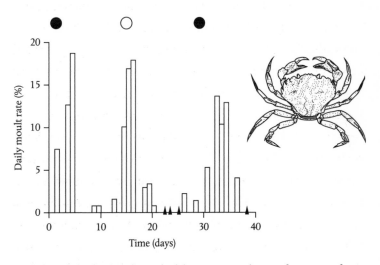

Fig. 11. Juvenile shore crabs in the laboratory moult just after times of new and full Moon.

such a vulnerable state, to be immersed in sea water, that is, during the highest spring tides. At other times of the monthly tidal cycle, particularly during the smallest high tides at neaps, the rising tide may not quite reach them. During such times, juvenile crabs remain hidden in moist conditions beneath boulders on the beach and they do not moult. Moulting during times when they are not reached by the tide would be disadvantageous for their survival, since they would not gain bodily support by being immersed in sea water. So, young crabs that have newly settled out from the plankton and are living at high levels of the shore show a fortnightly pattern of moulting that is in phase with times of the full and new Moon, partly under the control of internal biological clockwork of semilunar periodicity. The possession of such internal clockwork clearly indicates a close association with lunar cycles, albeit through the intermediate effects of the Moon on ocean tides.

In contrast, a common sandhopper of European beaches, *Talitrus saltator*, shows a fortnightly pattern of moulting that is the converse of that of green crabs. These animals, which are active on the tidal beach at low tide, burrow near high-water mark at high tide. They moult most frequently around the times of the quarter Moons, more or less coincident with neap tides, when their burrows are least likely to be flooded by the high tides of springs.[70] Unlike crabs, immersion in the sea would be disadvantageous to a moulting *Talitrus*. Sandhoppers are essentially semi-terrestrial animals whose survival depends upon avoiding the frequent immersion in sea water that is a requirement for most shore-living animals and plants. Again, though, like the crabs the moulting rhythm of sand-hoppers is intimately related to the timing of tides, but is ultimately phased to the lunar cycle to the extent that it too is probably driven by biological clockwork of semilunar periodicity that the sandhoppers have acquired throughout the course of evolution. The question remains, however, as to whether the effects of lunar cycles are so pervasive that it has been possible for living organisms to have acquired Moon-based rhythmicity without being exposed to the intermediate environmental factors associated with Moon-driven tides.

* * *

In searching for possible Moon-related biological rhythms in marine animals living in localities that are not influenced by tides it is initially tempting to consider that species living away from tidal coasts would be worthy of investigation. Recent discoveries have established that several deep sea animals exhibit apparent lunar rhythms of reproduction.[127] However, animals living in inshore waters away from the direct influence of tidal rise and fall along a coastline are also not immune from tidal cues. Sensitivity to tidal rise and fall is certainly known to occur in some bottom-living fish in the open sea, such as, for example, the commercially important plaice *Pleuronectes platessa*, which may not be exposed to the air at low tide. Larval stages of this flatfish, once they are shed to swim in the plankton above plaice spawning grounds, tend to drift away in tidal currents, eventually settling out on the sea bed as typical flatfish some distance away from where they were born. Yet by the time they are mature they must return to the original spawning ground again, raising the question of how this 'homing behaviour' is achieved. The problem has been solved by capturing plaice, fitting them with tiny depth recorders that do not affect their ability to swim effectively, returning them to the sea, and following their movements using tracking devices aboard a research ship.[71] By this means it was found that when the fish make the return journey to the spawning grounds they swim up from the sea bed when the tide is flooding, carrying them towards their objective, but sink back to the bottom, staying in position there when the ebbing tide would tend to take them in the reverse direction. In this way they use tidal currents to take them in the direction of their spawning grounds by selectively swimming off the sea bed at times determined by their internal biological clocks of tidal periodicity. Even away from the coastline, therefore, with no tidal cues from the crashing of waves on the shore, or periodic exposure to the air at low tide, the young plaice must perceive pressure changes associated with tidal rise and fall or be sensitive to changing patterns of tidal currents. If plaice, living at some depths in the near-shore are able to exhibit Moon-related behaviour in this way then it would not be surprising if other fish species too were sensitive to Moon-related phenomena.

Certainly it is known that Moon-related rhythms of spawning that match the fortnightly pattern of new and full Moon high spring tides are apparent in a wide range of inshore-living fishes. Such breeding patterns have been reported in Atlantic silverside, tropical coral reef saddle back wrasse, tropical damsel fish, surf smelt, four-eyed Amazonian fish, and a southern hemisphere galaxid fish. The last of these is a river-living fish whose larvae undergo their early stages of development in the sea. Adult galaxids spawn among vegetation on the banks of estuaries at times that permit newly hatching larvae to swim free when ebbing high spring tides in their native estuary carry the young fish seawards. Whether the spawning rhythms of such fishes are primarily driven by direct perception of the lunar cycle or indirectly through the influence of fortnightly spring tides is not clear, but these two recurring questions have been addressed in detail by biologists concerned with the breeding behaviour of some other fishes.

Particular attention has been paid to two species of killifish, *Fundulis grandis* and *F. heteroclitus*, the former found in the Gulf of Mexico and the latter, also called the mummichog, found on the Atlantic coasts of the eastern USA.[72] Mummichog that live in salt marshes of the north-eastern coasts of Florida begin their breeding season early in the calendar year, first by spawning during a succession of full Moon periods, that is, with pronounced lunar monthly periodicity, as if in response to times of maximum moonlight. Surprisingly they then continue to spawn until later in the year when their pattern of egg release changes. After spawning for some months at times of the full Moon, they change to less intense bouts of spawning around the times of both full and new Moon, that is, at times of maximum and minimum moonlight intensity. It is as if they switch from spawning in response to times of maximum moonlight to times of high spring tides, before ceasing to spawn in September.

In contrast, Gulf of Mexico killifish exhibit spawning behaviour that is at all times synchronized only with the semilunar pattern of high spring tides, with no evidence of a monthly pattern associated directly with the lunar cycle. However, in both Gulf killifish and Atlantic killifish, the fish persist with approximately fortnightly rhythms of spawning when they are kept in constant conditions in the laboratory, with no exposure to the Moon or to changes in the depth of water in which they are kept. Like several other animals that have been discussed so far, it is evident that over evolutionary time killifish have acquired internal biological clocks of approximately semilunar periodicity. This biological clockwork allows the fish to set their times of spawning to coincide either with the phases of the Moon directly or with the associated indirect Moon-driven changes of tidal height. Killifish in the Gulf of Mexico appear to rely solely on the cycle of high spring tides to synchronize their reproductive cycles, but there is circumstantial and additional experimental evidence that moonlight does directly synchronize the reproductive rhythm of Atlantic coast killifish, at least in the early months of the year. Atlantic killifish in laboratory aquaria which are exposed to cycles of artificial moonlight have been shown to entrain to a monthly reproductive cycle similar to that which they exhibit in nature in the early months of the year. Here, the evidence suggests that the distinction between so-called indirect effects and direct effects of the Moon is becoming blurred. However, each type of effect supports the notion that Moon-related biorhythms in some killifish, as in many other animals referred to so far, are causally related to the physical cycles of the Moon in relation to the Earth. Also, as mentioned earlier, some killifish switch their times of maximum spawning from new Moon high tides to full Moon high tides and back again, following the fourteen-month syzygy inequality cycle of lunar tides.

Even so, among many marine fishes it is hard to find unequivocal evidence that they show responsiveness directly to the Moon. There is anecdotal evidence that the coelacanth off the Comoros Islands in the Indian Ocean is more venturesome during full Moon phases. However, it cannot be excluded that, at the depths at which they are found (the first living specimen was trawled in about 40 fathoms), coelacanths may be influenced by the increased strength of tidal currents during high spring tides and not by moonlight directly. Accordingly, a search for Moon-related behaviour in fishes might be more successfully pursued by studying the breeding cycles of fish living away from the influence of tides, very deep in the sea or in rivers and lakes.

Deep ocean-living fish that have been studied in this regard are lantern fish in the north Atlantic. Several species of these fishes have been observed to undertake substantial upwards and downwards migrations in the open sea, and in two species, *Hygophum macrochir* and *H. taaningi*, the migrations have strong correlations with the lunar cycle,[73] probably with little influence from tides. Remarkably the fish also have their vertical migration patterns coded into their otoliths—bones in the ears of fish that are classically used to determine their age. In areas of the globe north or south of the equator where there is strong seasonality throughout the calendar year, periods of intense feeding and strong growth in spring are reflected in characteristic growth rings in the otoliths, like those laid down in trees. In lantern fish, however, when otoliths are sectioned they show additional growth rings that also require interpretation. The fish have been caught using nets towed at precisely determined depths at various times throughout the lunar cycle. By this means it has been shown from catches taken during nights of the full Moon that some lantern fish migrate upwards in the ocean from depths greater than 400 metres to depths shallower than 100 metres, where they feed intensively. In contrast, during the new Moon and other phases of the lunar cycle they tend to remain in cold, deeper water where feeding is less intensive. After they have been feeding in warm, near-surface waters a characteristic growth ring can be discerned microscopically in the otoliths. Such rings are recognizable as representing periods of fast growth, and they alternate with bands of less easily distinguishable structure associated with periods of slow growth in deep, cold waters where, by and large, the fish remain during periods of the new Moon. The otoliths therefore serve as 'data-loggers' that carry a record of vertical migration by lantern fish and show that upwards migration into surface waters to feed is clearly related to the lunar cycle. If by any chance the fish perceived full Moon high spring tidal currents at depths of 400 metres one would expect them also to perceive similar currents during new Moon high spring tides, but they clearly remained at depth during new Moon. Conclusively, therefore, it appears that upwards migration during full Moon is in direct response to the intensity of moonlight during that phase of the lunar cycle, and not to the indirect effect of tides.

Other fish that are not normally influenced by tides are those living in rivers. If any of these fish were found to exhibit Moon-related behaviour one might expect their patterns of behaviour to be truly monthly in direct response to the presence or absence of moonlight. Certainly one would not expect such fish to exhibit the fortnightly patterns of behaviour that, in marine species, might be related to the secondary effect of Moon-driven tides. But semilunar biorhythms have indeed been observed in some river fishes, raising the question as to how they have evolved. For example, juvenile rainbow trout in Canada have been shown to exhibit fortnightly cycles of growth, with greatest growth spurts occurring four to five days before full and new Moon.[74] This fourteen-day growth pattern is evident even in rainbow trout that are grown in laboratory tanks illuminated only by an artificial light–dark cycle of twelve hours light and twelve hours dark. The semilunar growth rhythm is therefore an inbuilt and inherited capability that probably has its origins in the evolutionary history of the fish. It is a river fish that has some tolerance of sea water, and some populations of rainbow trout still migrate between the sea and the rivers in which they spawn. The semilunar rhythm of growth in rainbow trout may therefore be a legacy of previous links with the Moon-driven spring–neap cycle of tidal seas. However, without repeated cueing of this semilunar biorhythm by exposure to monthly tides, it must be assumed that synchronization of the rhythm is in direct response to the lunar cycle. There is other evidence too that moonlight directly affects the behaviour of some river fish. Juvenile Atlantic salmon in Scottish rivers, that have not yet experienced the influence of tides in the sea towards which they are heading, have been observed to show intermittent downstream movements in their seaward migration pattern, most strongly during the new Moon phase of the lunar cycle. These fish, too, must be assumed to possess inherited biological clock capability of approximately monthly periodicity.

But perhaps the most conclusive evidence of behaviour patterns in aquatic animals that are directly controlled by the Moon is to be found in animals that inhabit land-locked lakes. Such water bodies have a constancy and virtual absence of tides which would lead to the conclusion that if animals in such localities exhibited Moon-related behaviour, then that behaviour should be under direct control of the Moon, presumably through the influence of varying intensity of moonlight.

There is certainly some anecdotal evidence that this is so, dating back to the early and mid-twentieth century. Notably it has been reported that in some Asian and African lakes the larvae of certain aquatic insects emerge as flying adults with lunar periodicity. The phenomenon has been widely observed in the mayfly *Povilla adusta* in lakes in Uganda, in which emergence of the flying adults is closely synchronized and occurs shortly after full Moon. In addition, swarms of 'lake-flies' have been observed to emerge from their larval habitats on lake shores shortly after new Moon.[75]

After such early and casual observations of Moon-related behaviour in lake animals, the first comprehensive description of the phenomenon of lunar spawning synchronicity in a lake-living fish was reported late in the twentieth century.[76] The fish concerned was a cichlid, *Neolamprologus moorii*, known only from Lake Tanganyika, where it inhabits shallow rocky areas at the southern end of the lake in Zambia. Direct evidence of spawning by the fish was difficult to obtain because eggs are normally concealed within crevices at the spawning site within the feeding territory of the adult stock. However, it was possible to estimate the spawning times by capturing larvae as they hatched as free-swimming fry, which remained in tight schools that hovered near the spawning locality. The larvae could then be observed in laboratory aquaria, during which time their growth patterns could be carefully measured over successive days. It was then possible by back-calculation from the growth curves of each brood to show clearly that spawning must always have occurred on or near the day of the first quarter of the Moon. As in marine animals, synchronized release of eggs and sperm has clear advantages for survival of the species, but the biological advantages of spawning at a particular phase of the Moon are less easily explained in lake-living than in marine species. In marine species, release of larvae at particular phases of the Moon can be explained in relation to the advantages of wide dispersal in the sea or retention near the parent stock, according to the associated timing of tides. In the absence of tides, the advantages to fish and other organisms of Moon-related spawning are less clear, but for a cichlid fish in Lake Tanganyika an explanation for precisely timed lunar spawning may be forthcoming.

In *Neolampralogus moorii*, eggs and young fish are concentrated in a small area within the feeding range of the adults and are potentially susceptible to predators. Accordingly, during daylight hours, the territory and the offspring are vigorously defended by the adult fish. In darkness, however, the eggs and young fish are susceptible to predation by catfish, which are olfactory or tactile feeders that can approach the spawning area without being seen by the adult cichlids. Is it possible therefore that the timing of egg-laying at the time of the first quarter of the Moon has advantages in reducing the risk of predation of larvae at night? Certainly, newly hatched and particularly vulnerable larval stages are abundant in the days up to the time of full Moon, following egg-laying at the time of the first lunar quarter. Also, during the period of the lunar cycle from the first quarter until full Moon it is just as the Sun sets that the Moon rises, illuminating the spawning area particularly effectively at night. The timing of larval hatching therefore occurs when moonlight is of greatest intensity and duration. This illumination at night assists the parental fish in seeking out potential predators and is also a deterrent to night-feeding predatory fish that prefer to hunt by touch and by olfaction. Indeed, hunting activity by nocturnal predatory fish is known to be suppressed during bright moonlight, even to local fishermen, who do not fish at night during the full Moon period. It can be argued, therefore, that brood survival is considerably enhanced by the timing of egg-laying to the days of the first quarter of the Moon. And there is additional evidence that this explanation of lunar periodicity of spawning is correct.[76]

Comparisons have been made between the brood survival rates of N. *moorii* and another Tanganyikan cichlid species, both of which show lunar periodicity of spawning, with the survival rates of two other species from the same lake which do not spawn with lunar periodicity. The former two species showed clearly enhanced early survival rates, presumably due to protection by the parent fish during moonlit nights, when compared with the latter two species. Here, then, we have examples which demonstrate quite convincingly how a biological rhythm keyed to a particular phase of the lunar cycle may have selective advantage and be of evolutionary benefit for survival of the species.

Coincidentally, too, improved understanding of the behaviour of African lake cichlid fishes also suggests other benefits of synchronized spawning, beyond the increased chances of fertilization by spawning *en masse*. First, the coming together of large numbers of breeding males and females may reduce the predation rate on the parent population through safety in numbers. Secondly, and perhaps more importantly, the mass release of larvae would most likely 'swamp' their potential predators, which would only be able to catch a limited number of prey at any one time.[68]

So, Moon-related synchronized spawning occurs in animals in the sea and in lakes, confirming that the phenomenon is not always mediated through the intermediate effects of Moon-driven tides. Many examples are now known of breeding rhythms and other patterns of animal behaviour that have evolved directly in relation to monthly changes in the intensity of moonlight. Cichlid fishes provide at least one example of the selective advantage of direct Moon-phased behaviour and all that remains is to consider how indirect Moon-phased behaviour might have evolved in response to Moon-driven tides.

A study that perhaps offers some clues to the process of adaptation to Moon-related tidal phenomena in the sea concerns the breeding pattern of bluehead wrasse, *Thalassoma bifasciatum*, a common coral-reef fish found throughout the Caribbean.[77] The adults of this fish spawn on the coral reefs, after which the newly hatched larval fish are distributed widely in the open sea. Eventually the larvae return to the reefs where they were hatched and find patches of sand into which to burrow. There, after a few days they metamorphose into fully formed juvenile fish which emerge from the sand and recruit back into the parent population. This recruitment pattern was found to be cyclical, most juvenile fish appearing on the parental reefs during neap tides, more or less coincident with times of the first and third lunar quarters. Surprisingly this fortnightly pattern of recruitment of the juveniles was unrelated to the pattern of spawning behaviour of the adults, which was not cyclical.

So, if the fortnightly pattern of recruitment was not related to a similar pattern of spawning, how was the recruitment pattern determined? The best explanation available is that it has evolved against the background of 'ocean weather', namely, the conditions whereby strong tidal currents flow around the reefs during fortnightly spring tides, alternating with weaker tidal flows during the intervening neap tides. It is not unreasonable to assume that the bluehead wrasse larvae recruit back to the parental reefs where they were spawned during the gentle tides associated with neaps, avoiding wide dispersal to unfavourable habitats away from the parental reefs by strong spring tides. The study therefore provides encouraging evidence in support of the notion that Moon-related cyclical behaviour has evolutionary advantages for bluehead wrasse, adding weight to earlier speculation that this is so in some invertebrate species. Furthermore, taken also in the context of the examples of cichlid spawning in non-tidal lakes, we can conclude that biological rhythms of lunar periodicity have advanced from the realms of anecdote to the world of testable reality. There is clearly increasing evidence of the validity of such rhythms, not only indirectly related to the Moon through the influence of tides, but also directly related to the Moon in the absence of tides.

Now we must consider wider implications of the role of the Moon in controlling animal behaviour in appearing to induce species interactions and patterns of feeding behaviour whereby one species avoids another. Lunar cycles are also implicated in phasing longer cyclical patterns of breeding than we have considered so far. At the same time we must also begin to extend the discussion of Moon-related phenomena from animals that live in aquatic environments to those that live on land. These are topics for discussion in the next chapter.

7

MOONLIGHT AVOIDANCE AND MOON COUNTING

Since historical times, human beings have had an innate fear of darkness and the changing position and appearance of the Moon in the night sky. There are many early references to an 'evil Moon'. From Charles Darwin onwards, many people have considered this to be the legacy of a primitive evolutionary adaptation in early human beings to the risks of attack by nocturnal predators such as big cats. The fear may have been rationalized away in sophisticated societies exposed to continuous urban illumination at night, but in traditional societies the bright nights of the full Moon often continue to be coupled with a fear of the night. So much so that many such societies still have myths and superstitions associated with times of the full Moon. The image of the silhouette of a wolf howling towards a full Moon reflects such beliefs. In some parts of Africa, villagers perceive the full Moon to be the harbinger of increased risk of attack by lions. As a sub-editor wrote above an article on this subject in the London *Times* in 2011: 'Beware a full Moon, it's a sign that you may soon be a lion's lunch.'

There is as yet little hard evidence that the behaviour of human beings is in any way directly regulated by the lunar cycle, except perhaps concerning sleep patterns, which are discussed later, but some persisting myths may have origins in the manner in which humans have come to respond to the behaviour of other animals. This raises the question of whether there is any evidence of animals responding directly to moonlight in such a way as to influence other animals, including humans, to exhibit related behaviour that therefore appears to be linked to the light of the Moon, albeit indirectly.

To consider this question in evolutionary terms we can look first at the manner in which different populations of animals, say predator and prey species, interact in the primitive environment of the sea. It is now a common occurrence in the world's seas and oceans, and probably was so in the Earth's early geological history, for vast quantities of planktonic animals to swim up from the dark depths to spend the night near the surface before sinking again at dawn, thus exhibiting diurnal vertical migration or DVM. It is well known too for modern-day planktonic animals that have migrated upwards at dusk to temporarily sink around midnight before rising again, ahead of their major descent at dawn. In some cases, at least, the temporary sinking in the middle of the night is thought to be due to the influence of moonlight, particularly during full Moon.[78] Measurements of light intensity in the open ocean during a full Moon show that any vertically migrating planktonic animals rising to within about 150 metres of the sea surface would be exposed to moonlight, a light intensity which is known to depress the upper limit of migration of some crustacean species on a monthly basis. Also, species of mid-water ocean prawns caught rising from depths of 400–500 metres to about 100 metres below the surface have been observed to sink temporarily to 160 metres depth at moonrise. So how does this supposed avoidance of moonlight come about?

Food chains in the sea begin substantially with the production of huge quantities of single-celled plants which require light for photosynthesis and growth, and which are accordingly confined to surface waters. As single cells these plants have little active mobility, forming green, near-surface pastures which deeper-living planktonic animals swim up to and graze upon. Upward swimming of the herbivorous grazers occurs at night, with the advantage that, at such times, they themselves are less likely to be preyed upon by carnivores that feed at or near the surface, including fish and birds, most of which hunt for prey by sight. That is not to say that the herbivores avoid being preyed upon at night, however, as some predators which feed upon them at depth by day follow them into surface waters by night. So, with a whole community of herbivorous grazers and carnivorous feeders rising into near-surface waters at night it is perhaps not surprising to find that the occurrence of bright moonlight around midnight results in the temporary descent again of some of the upward migrants. It would be advantageous for the grazers to lead that descent to avoid being seen and fed upon in bright moonlight. It might of course be equally advantageous for some of the near-surface-feeding carnivores to follow the greatest density of their herbivorous prey during their short and temporary descent, despite the increased difficulty of finding them in the darker depths. In fact, though, not all the predators, including some fish, do descend during the middle of the night, suggesting that there are disadvantages to them if they follow their prey at that time of night. The explanation is perhaps that fish often capture their prey from below, by seeing them in silhouette against the lighter surface of the sea above.[79] This is a method of capture which would be unreliable during night-time hours when the prey species had moved deeper in the sea, below moonlit surface waters.

There may be other factors affecting the downward migration of plankton in the middle of the night, particularly since it can also occur on nights of the new Moon when moonlight intensity is low. Nevertheless, the consensus among oceanographers is that avoidance of moonlight could explain some of the examples of 'midnight sinking' that are well known to occur in ocean plankton.[27] Indeed, moonlight is also known to affect the behaviour of some bottom-living species that live in inshore waters, many of which have been observed to rise from sand or mud to swim upwards to join planktonic species in their nocturnal swim towards the surface at night. Like planktonic animals, many of these species, too, are known to be deterred from swimming upwards off the sea-bed in bright moonlight. During full Moon periods, some even delay their upwards swim until moonset or return to the sea-bed at moonrise, clearly showing evidence of avoidance of bright moonlight.[80]

Even Galapagos fur seals (*Arctocephalus galapagoensis*) are known to avoid bright moonlight in the sea. About twice as many fur seals have been observed to be ashore at full Moon than at new Moon,[81] possibly because of the behaviour of their food in the plankton or, perhaps more likely, to improve their chances of avoiding predation by night-hunting sharks. It has also been reported[82] that the ocean-skater *Halobates* is able to avoid capture by tow-nets during the light of the full Moon. The ocean-skater is one of the few species of insect found in or on the sea and is a member of the neuston, those animals that reside at or on the ocean surface. Special nets towed from ships are used to sample the neuston, and oceanographers at first discovered unaccounted for variability in samples of *Halobates*, which is in any case patchily distributed in time and space. The variability was explained when sampling times in relation to lunar phase were taken into account. It was concluded that the insects were able to dodge neuston nets deployed from a research vessel, by the light of the full Moon. Therefore moonlight seems to have a marked effect in inducing avoidance behaviour in many invertebrate and vertebrate animals that live in or even on the sea.

Moonlight also seems to influence many animals that live in fresh waters and on land. In rivers of the USA a species of crayfish (*Orconectes virilis*) is known to be most active at new Moon and least active at full Moon, even when the sky is overcast.[83] The most reasonable explanation for this difference is that it reduces their chances of being eaten, since most predators of crayfish, including ducks, herons, fish, turtles, raccoons, muskrats, and humans, rely on sight to capture their prey. Intriguingly, since the crayfish recognize phases of the Moon, even when the sky is overcast, they appear not to be simply responding to the intensity of moonlight in their Moon-related pattern of foraging behaviour. Perhaps they, like many other animals, possess internal biological clocks of lunar periodicity that also help to keep them safer than they would be if they were out in the open and visible to predators at all times of the lunar month. Such behaviour, under the control of internal lunar clockwork, that has been acquired and inherited over evolutionary time, would further testify to Moon-related behaviour's evolutionary importance.

Yet more evidence demonstrating increasing acceptance of such a conclusion, this time with reference to terrestrial animals, was published in the prestigious *Proceedings of the Royal Society* in 2010, under the title 'Moonlight avoidance in gerbils reveals a sophisticated interplay among time allocation, vigilance and state-dependent foraging'.[84] This and many other published studies provide credible evidence in support of the notion that moonlight avoidance occurs particularly commonly among rodents, including gerbils, kangaroo rats, pocket mice, porcupines, wood rats, spiny mice, and deer mice. Banner-tailed kangaroo rats, for example, that are strictly nocturnal in winter, avoid foraging above ground at night when the Moon is full.[85] On bright moonlit nights they tend not to venture above ground, thus reducing the risk of being captured by predators. Essentially all these animals undertake a type of risk management: they balance their hunger state and their need for food against the risks of exposing themselves to predators. In other words, they make trade-offs between feeding and safety. Such animals normally feed at night, which reduces the risk of being eaten by predators that hunt by day, and they reduce their risk even more by avoiding bright moonlit nights. In these circumstances it turns out that even the extent to which nocturnal rodents exhibit an awareness of predators is evolutionarily programmed to vary cyclically according to the phase of the Moon, with greatest awareness of potential threats from predators occurring during times of full Moon.

In fact the interaction between rodents and some night-time predators can be likened to a 'biological arms race'. As night hunters that seek their food by sight have adapted by increasing the efficiency of food capture during bright moonlight, their prey species have responded by tending to avoid moonlight as they forage for food. In true 'arms race' style, as night hunters such as owls and foxes increase their ability to capture rodent prey by moonlight, so the rodents increase their ability to avoid moonlight accordingly.

Other predatory animals, too, are able to phase their feeding patterns to the light of the Moon. One of these is the ant-lion (*Myrmeleon obscurus*), a dragonfly-like insect whose larvae are also able to distinguish between the light intensities of full and new Moon. The eggs of this species of ant-lion are laid in sand and each larva, as it emerges from the egg, excavates a funnel at the sand surface, burying itself at the base of the funnel with its large jaws projecting upwards as it lies in wait for food. The trap is set to capture ants and other insects as they fall into the funnel where they are pierced and sucked dry by the growing ant-lion larva. Intriguingly, the larvae excavate larger funnels during times of the full Moon than at other stages of the Moon,[86] presumably at times when their prey are less abundant or more wary of life's pitfalls (Fig. 12).

Moreover, a similar monthly rhythm of increase and decrease in the size of the funnels is exhibited by ant-lions kept in constant conditions in the laboratory. Here again we have an example of an animal that has not only adapted to the monthly cycle of moonlight but has acquired internal biological clockwork of circalunar periodicity to permit it to do so.

* * *

But moonlight-related behaviour extends even more widely among aquatic and terrestrial animals. African elephants, for example, where they range near sites of human occupation, can be seen to have refined their night-time foraging on crop plantations to nights when moonlight is at a minimum, presumably reducing the risk that they will be discovered and

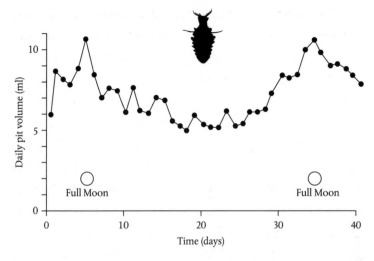

Fig. 12. The ant-lion and the Moon. An insect which builds pitfall traps in sand to catch prey, constructing larger pits during full Moon than at other times of the lunar cycle.

driven off by farmers. Avoidance of moonlight is even seasonal in the snowshoe hare of North America,[87] whose avoidance varies according to changes in moonlight intensity which are associated with the amount of snowfall throughout the year. During the winter snow season, hares were found to be more secretive during the particularly bright moonlit nights of full Moon than at other phases of the Moon. In contrast hares were indifferent to changes of moonlight intensity throughout the lunar month in the snow-free season, when moonlight was not reflected by snow cover.

Moon-related behaviour also occurs in some marine birds that feed at night, such as petrels which are themselves prey for some large predatory birds such as skuas. This is an interplay which has prompted biologists to ask whether changes in moonlight affect patterns of feeding behaviour in either the petrels or the skuas, or both.[88] Unlike most seabirds which feed by day, the vast majority of small petrels are nocturnal in their habits as they feed at sea in the vicinity of their coastal breeding grounds, hunting for near-surface planktonic animals, many of which are luminescent and can be caught by sight. Certainly, some petrels are known to reduce their food-searching activity considerably during bright moonlit nights. At such times, it could be argued, moonlight avoidance comes about because some of the planktonic animals upon which the petrels feed, particularly crustaceans, temporarily descend from the sea surface during the middle of the night. But there would be added advantages to petrels in avoiding feeding in large congregations at the sea surface during full Moon; in doing so they would also reduce the risks of being preyed upon themselves by larger, night-feeding skuas. Further anecdotal evidence that some petrels use moonlight-avoidance behaviour to their advantage is that some of them even tend to avoid congregating near their colony roosting sites in bright moonlight, where in dense concentrations they would also be at high risk of capture by night-hunting skuas. In other words, even to the casual observer, petrels tend to avoid moonlight during full Moon periods, with conflicting explanations of the advantages to them of doing so. The question for ornithologists was: do petrels disperse during bright moonlight because of a supposedly reduced ability to find food, or do they do so in a manner which reduces the risk of being preyed upon by night-hunting skuas?

Attempts to resolve which of these two possible explanations was the most likely were made by observing the behaviour of populations of blue petrels (*Halobaena caerula*), which are preyed upon by brown skuas (*Catharacta antarctica*), both of which nest on the Kerguelan Islands in the Southern Ocean.[88] First it was observed that brown skuas caught more prey at night than during the day, with greatest catches of petrels by skuas occurring during times of full Moon. The risk of capture of petrels at these times was found to be greater in non-breeding birds than in members of the breeding population, suggesting that wide dispersal of breeding adults on moonlit nights is primarily advantageous in avoiding the attentions of the skuas. It was also found that it was the breeding population of petrels that avoids hunting for food at the sea surface during moonlit nights. Evidently, by the time they are adult the petrels have learned that it is of great advantage to them in avoiding the attentions of skuas if they do not congregate near their nest sites or in feeding flocks near the sea surface during times of the full Moon. On the Kerguelan Islands, therefore, whole colonies of seabirds, including predators and prey, are partly governed in their daily and monthly patterns of behaviour by cyclical changes in the light of the Moon.

* * *

So if, as seems likely, avoidance of moonlight due to the risk of being eaten is such a prevalent form of behaviour in the animal kingdom, what might this tell us about deep-seated fears in many people throughout the world in relation to the night and to the Moon? A fear of lions is a long-standing characteristic of human communities and we may ask if historical associations between human beings and their potential predators, such as lions, have any linkages to the lunar cycle. Lions of the African plains are primarily nocturnal feeders and, like many animal predators that we have discussed so far, are similarly patterned in their feeding behaviour in relation to the Moon. They are least likely to be able to capture herbivorous animal prey during moonlit nights of the full Moon, because of the greater awareness at those times by their prey. More than this, though, it is known that the majority of lion attacks on humans, which also tend to occur at night, vary in their intensity according to the phase of the Moon.

Studies have been conducted using meticulous records of the timing of lion attacks on over a thousand people in Tanzania over a period of several years.[89] The investigation was supplemented by observations of the hunting, feeding, or resting state of lions located in the Serengeti and Ngorongoro National Parks after fitting them with collars to which radio-transmitters were attached. The studies showed that lions were least dangerous to humans during times of full Moon, again presumably because of greater human awareness at those times, but they were most dangerous to humans in the first weeks following the full Moon when the Moon is below the horizon. Most lion attacks on humans were found to have taken place between dusk and around 10 p.m. during the waning phase of the Moon when it rises well after dusk, which in the tropics is consistently about 7 p.m. The most dangerous times for humans living in lion country are therefore nights a little after times of the full Moon, usually just after sunset. Unfortunately this is the time when people tend to become active after the heat of the day, the very times of the month when early evenings are at their darkest. Unlike some stories of folklore, then, it turns out that people are safest from lion attack during well-lit nights of the full Moon, but the risk to humans of predation by lions increases markedly in the evenings following the full Moon. Hence the newspaper headline referred to at the beginning of this chapter.

Homo sapiens and his early relatives have always lived in close proximity, not only to lions, but also to other big cats such as leopards, tigers, and jaguars, and large, nocturnal predators such as wolves, all of which may be at their most dangerous during times of the waning Moon. It has been reported that lions were once the most widely distributed mammal through-out the world and big cats are still found co-existing with native human populations throughout the tropics in Asia and South America as well as in Africa. Indeed, early humans apparently scavenged from the kills of big cats, and lions were painted in lifelike detail in cave art 36,000 years ago. This confirms a long-established historical association, indicating that man has always been exposed to the risks of predation that would have cycled with the waxing and waning of the

Moon during the millennia of human existence. Moreover, man's own cyclical behaviour appears to interact with the Moon-related hunting behaviour of lions to determine the precise pattern of human susceptibility to lion attack, even to the present day. Grazing animals, that are also preyed upon by lions, range freely at night and are highly vulnerable to lion attack throughout most of the lunar cycle, apart from during full Moon when grazing herbivores would have a greater chance of avoiding big cat predators. In contrast, humans have tended to forage mainly after the heat of the day, seeking shelter for sleep much later in the evening until dawn, behaviour which has led to increased vulnerability to lion attack in the early part of the night, at times signalled by the waning of the full Moon. Little wonder then that the full Moon as a portent of dangers to come is deep-seated in the psyche of some human societies. The legacy of Hippocrates, who considered that moonlight caused nightmares, persists in some people, and the fear of darkness, with associated Moon myths, remains in many societies to the present day.

But Moon-avoidance is not a universal phenomenon among members of the animal kingdom. One species of bird at least, the wideawake tern, seems to have learned how to recognize occurrences of the full Moon and, moreover, seems to be able to count them off against the timing of its own physiological breeding cycle.

* * *

Wideawake terns are so called not because of their awareness of the lunar cycle, but because of the sound of their call, 'wide awake', and no doubt also because of the sleepless nights endured by humans living close to their raucous breeding colonies. They are sooty terns (*Sterna fuscata*), marine birds which defy a common assumption that the breeding pattern of birds is annual. In one of their breeding locations, at least to the casual observer, the breeding times of sooty terns might appear to be quite random throughout the year, the precise times of nesting and egg-laying unpredictable from the annual solar calendar. On the equatorial Atlantic Island of Ascension the breeding sites of this bird may be noisily occupied at any time of the solar year, times known as 'Wideawake Fairs' which can be predicted only by experience and by consultation of a lunar calendar. The correlation between the breeding times of sooty terns on Ascension Island and the monthly cycles of the Moon is quite remarkable.

Sooty terns, with their black cap, back, and legs, and white underparts, are larger than black terns, which they superficially resemble, and are most commonly seen throughout the world's tropical seas. Only rarely are they seen by observers of the seas of northern or southern latitudes, and only then as vagrants when they are displaced from their normal locations by strong winds. Indeed, they rarely come to land except to breed, and may be at least six years old before they do so. Their feeding therefore has to be at sea, where they take fish, including flying fish, squid, and planktonic crustaceans which may be driven towards the ocean surface by larger fish feeding below. When they do come ashore to breed they do so most commonly on rocky and coral islands throughout the equatorial zone. Whether at sea or on land they are highly gregarious and occur in very large flocks in isolated localities, always returning to the same nesting sites to breed. As a result of their loyalty to a particular breeding location it is possible to recognize sub-species or varieties of sooty terns associated with their nesting sites on different oceanic islands.

A colony nesting on Bird Island in the Seychelles in the Indian Ocean numbered more than a million birds at its peak, and an uninhabited island in the Kiribati Islands, just south of the equator in the Pacific Ocean, once boasted a colony of 1.5 million pairs. The birds nest among rocks and coral rubble in relatively level locations, usually avoiding cliff faces. Accordingly, and not surprisingly, the Kiribati location was the site of extensive guano exploitation in the nineteenth century. Colonies on the Hawaiian Islands are referred to as 'ewa ewa', which roughly translates as 'cacophony', again reflecting the incessant calls of the birds when in residence at their breeding sites.

On Ascension Island two particularly notable colonies of sooty terns are to be found, in company with a number of other nesting seabirds. The two nesting colonies, or 'fairs', estimated at the beginning of the twentieth century to number up to 400,000 birds, are well established in what are referred to as the Mars Bay Fair and the Waterside Bay Fair, both in the south-west corner of the island. The two colonies have been carefully observed by ornithologists since the mid-twentieth century and, throughout that time, regular records have been kept each year of the times of arrival of adult birds at their nesting sites after their times spent at sea, beginning with nearly twenty years of records published in 1959.[90] Accordingly it has been possible to document over a number of years the precise times of arrival of the first birds for nesting and the times when the first eggs were laid. Initially, even with hard data being collected, it seemed that the nesting and egg-laying season occurred quite randomly throughout the year—until sufficient records were obtained to show otherwise. Eventually it turned out that a 'Wideawake Calendar' could be formulated which showed that adult sooty terns returned to the island to establish their nest sites around the times of every tenth full Moon. There then followed a period of up to six weeks or so when the cacophony of noise around the colonies was such that the event came to be referred to as the 'Night Club' phase or 'Wideawake Fair', during which, on average, egg-laying began around five weeks after the first birds arrived (Fig. 13).

The timing of onset of the 'fairs' and subsequent timing of egg-laying are remarkable biological phenomena which, if documented only on the basis of historical, anecdotal records, would today be considered to be Moon-related myths, to be considered with great scepticism. However, records have been maintained by professional ornithologists into the twenty-first century and these confirm that the Moon-related pattern of breeding is persistent.[27] All the same, a seemingly infallible breeding cycle of ten lunar months, even based on modern

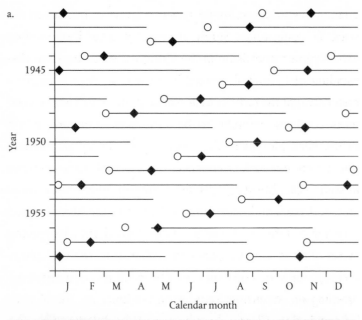

Fig. 13a and b. The Wideawake Calendar: wideawake (sooty) terns arrive to nest on Ascension Island in the mid-Atlantic shortly after every tenth full Moon (circles); the first eggs (diamonds) are laid within a few weeks of arrival and the terns leave after several months in residence.

quantitative evidence, might generate some scepticism in the scientific community. Surely, it could be argued, one would expect some imprecision in these timings, based on an understanding of biological variability. In that sense it is perhaps fortunate that the terns occasionally err and return after eleven months, adding credibility to the scientific basis of lunar timing of the fairs. Moreover, the credibility of the lunar breeding phenomenon is further enhanced since it is also the case that in some years breeding fails altogether. In other words, there are good years and occasional bad years in the Moon-related breeding pattern, exactly as occurs in 'normal' annual breeding cycles of birds and other animals such as fish. There is, nevertheless, sufficient constancy in observations of sooty terns breeding on Ascension Island for a ten lunar month cycle to be convincing. Clearly, breeding every twelfth or thirteenth lunar month would be a closer approximation to the solar year, consistent with an annual breeding pattern which one might intuitively expect. Amazingly, then, evolution by natural selection appears to have favoured a nearly perfect decimal system of counting lunar months and not the conventional annual system which counts twelve or thirteen such intervals of time.

As described in an earlier chapter for the timing of 'palolo days', the pattern of 'Wideawake Fairs' is consistent with the metonic cycle of nineteen years' periodicity. This is the cycle of progression of lunar months against solar years which ensures that if the day of the full Moon, and hence a fair, occurs, say, on

1 June in any one year, such a coincidence does not happen again until nineteen years later. But how is it that a ten lunar month cycle of breeding has come about when a twelve calendar month cycle associated with the solar year might be expected?

The answer to this question probably lies in the fact that, as mentioned earlier, sooty terns occur particularly abundantly in tropical regions, where they are subjected to smaller annual changes in day length and temperature than birds that are more commonly found in temperate regions of the world. In general, in temperate regions, birds mature on an approximately twelve-month cycle determined by their physiological condition. They breed truly annually, phased by seasonal changes in temperature and day length, and consequent seasonal changes in weather and vegetation in the environment. In tropical localities, particularly over the open ocean, away from many significant seasonal environmental cues, moonlight is perhaps a more perceptibly changing environmental cue. So if, as seems likely, sooty terns possess an innate physiological, approximately annual cycle of reproductive maturity, as do their relatives living in temperate localities, then the approximate nature of that cycle, signalled by lunar events rather than annual changes in day length, can be envisaged as accounting for the ten lunar month pattern of breeding. The tropics and sub-tropics are regions of anticyclonic weather conditions, where moonlight would be expected to be a more consistently perceived environmental factor than in higher latitudes where weather conditions are more typically cyclonic with more cloud cover. Sceptics may doubt whether the timing of Wideawake Fairs is determined directly by the Moon, but it is difficult to avoid the conclusion on scientific grounds that this is the case. In some examples of Moon-related biological rhythms that we have considered, the effect of the Moon has been seen to be indirect, through the intervention of the Moon-driven cycle of neap and spring tides, but responsiveness to the neaps–springs tidal cycle in this case seems unlikely.

* * *

A cue related to the lunar cycle, which may be relevant to both moonlight avoidance and Moon-counting behaviour, is that of the so-called Moon Illusion. This is a phenomenon whereby the perceived size of the Moon changes in a manner that is independent of apparent size differences associated with the Moon's proximity at various times during its elliptical orbit around the Earth. When the full Moon rises in the east, on the opposite side of the Earth from the setting Sun, it is generally observed to appear larger than when seen later at night when it is higher in the sky. The phenomenon is particularly striking at the time of the summer solstice when the high, apparent trajectory of the Sun is counter-pointed by a low, horizon-hugging full Moon. The illusion has been recognized as such at least since the time of Aristotle in the fourth century BC, but no doubt had mystical implications for human societies before and after that date. Along the way it has also served as the basis from which, in the seventeenth century AD, the French philosopher Malebranche considered that the Moon Illusion demonstrated the unreliability of the human senses. He argued that, because of it, man should distrust his senses and the so-called reasoned information which those senses provided. Consequently he proposed that man's search for the truth about life could not be achieved through his senses, but only through authority, namely that of the Church.[91] Later, the illusion was also used to challenge conventional views about the use of geometry in optics as a basis for understanding vision. Even into the early twenty-first century explanations for the phenomenon were still being sought.[92]

Greek scientists from Aristotle to Ptolemy, five centuries later, explained the Moon Illusion on the basis of atmospheric conditions near the horizon as viewed by the observer. Moisture and distant haze were proposed to act as a lens which enlarged the image of the rising Moon compared to the 'normal' image seen in the clearer atmosphere above. This interpretation persisted through later centuries, despite being challenged by eminent scientists such as Johannes Kepler in the seventeenth century, who argued that the proposed atmospheric lens effect could not possibly enlarge the Moon to the extent observed.

Today it is still sometimes presumed that the Moon Illusion is no more than an atmospheric phenomenon, but that is incorrect and more sophisticated explanations have been put forward, two of which have received particular attention in recent years. One of these proposes that the horizon full Moon is perceived as larger because the perceptual system, seeing the Moon near the finite horizon, treats it as though it is much farther away and compensates accordingly.[91] It was once thought that this interpretation depended on observation of the rising Moon behind trees or buildings on the horizon, but these are not necessary for the effect. The illusion is just as compelling when moonrise is observed above a calm sea, lake, or featureless desert. Another interpretation is based on evidence that when observers in an experiment viewed the rising full Moon through a narrow aperture, the image was seen to shrink in size and sharpen in appearance.[91] This suggests that, at least in part, the apparent enlargement of the Moon on the horizon is due to out-of-focus optics, coupled with enlargement of the pupil of the eye due to dim light intensity at dusk. If the Moon Illusion is apparent to sooty terns it would provide a consistent lunar monthly cue, enhancing the periodic experience of the full Moon itself, ten experiences of which might serve as the external timer for the birds' approximately annual internal rhythm of reproductive physiology.

* * *

Repeatedly throughout this book mention has been made of homing behaviour in animals, such as that exhibited by wide-awake terns in returning to their island nesting sites, and of the notion of biological clocks which reflect the heritability of adaptation by organisms to cyclical astronomical events. So far, however, we have not explored the extent to which these two kinds of phenomena might be connected. Little mention has been made of a feature of inherited biological clocks referred to by Klaus Hoffmann,[93] an expert in animal navigation, who wrote: 'One of the most spectacular aspects of biological clocks is their participation in celestial orientation.' This statement will be acknowledged in the next chapter as we consider the Moon's influence on homing behaviour by animals, after first discussing the history of Moon-related navigation procedures and associated advances in chronometry developed by early human mariners.

8

HOMING BY THE MOON

Modern navigation systems for ships and aircraft, or even for truck drivers finding themselves wedged in unfamiliar narrow roads, rely on global positioning information beamed from man-made satellites in orbit around the Earth. But it was not always so, and the problems of homeward navigation were particularly life-threatening for early ocean navigators venturing out of sight of land. For them the risks were serious until burgeoning interest in the apparent movements of the Sun, Moon, and stars began to yield superficial guidelines that helped them to orientate and navigate accordingly. At first they could obtain only sketchy information concerning latitude, the pole star indicating north in the northern hemisphere, and its height at night, or the height of the Sun at noon, giving some estimate of latitudinal position. Even the Moon at certain phases of the lunar month could be useful in indicating north or south, since a line drawn from the upper to the lower point of the crescent Moon indicates south in the northern hemisphere and north when south of the equator.

However, on the open ocean, estimation of latitude is only part of the problem—determination of longitude is much more complicated. Essentially, calculations depend upon recognition of the relationship between east–west displacement of the navigator and the passage of time in the context of the Earth's rotation. If a marine navigator knows the time at a fixed reference point, say the home port, and can compare that with local time, as determined by the observed apparent movement of the Sun, the time difference can be used to calculate the longitudinal distance from home. The calculation depends upon knowledge that the Earth rotates at 15 degrees per hour, generating longitudinal displacements that vary according to latitude, from a minimum at the pole to a maximum at the equator.

The first historically recorded recognition of the relationship between longitude and time was by Hipparchus, a Greek astronomer and mathematician living in the second century BC. Later the Islamic scholar Al-Burani appears to have been the first to report that the relationship between time and longitude was based on the knowledge that the Earth rotated on its axis. However, even with knowledge of the Earth's rotation, a critical requirement remained for early navigators. They still needed accurate estimates of the solar time difference between their new position and their home port. This problem is reported to have been solved by Amerigo Vespucci, one of the first world ocean explorers, during his first voyage to America in 1499, using a method based on 'lunar-occultations'.[94] Lunar occultations occur when the Moon obscures another celestial object such as a known star or planet, the Moon being chosen for this because of the speed of its orbital movement. For example, the Moon travels over ten times faster than the Sun in its observed celestial arc, a speed which allows for timings of its occultations to be recorded fairly accurately. One such conjunction, of the Moon with the planet Mars, occurred during Vespucci's voyage on 23 August 1499, a phenomenon that the explorer recorded in his diaries. At that time astronomy was sufficiently advanced for Vespucci to have available to him astronomical tables on the timing of lunar occlusions, which would be observable simultaneously at his home port in Europe and at his new, unknown locality. From the astronomical tables he would know the expected time of the lunar occultation of Mars as viewed from Europe, and could compare that timing with the time of day at his new location, which he could estimate from observation of the Sun's apparent movement in the sky. Hence the time difference between the home port and the new locality could be calculated to give, for the known latitude of the ship's position, at least an approximate estimate of the longitudinal distance between the home port and the present location. Vespucci would know that the Earth's rotation, that is, the apparent orbit of the Sun, takes place at approximately 15 degrees each hour. Accordingly, the total displacement angle could readily be converted to displaced distance appropriate for his latitude in nautical tables available to him.

More critical determination of longitude for mariners, less dependent upon knowledge of intermittent lunar occultations, had to await development, in the seventeenth and eighteenth centuries, of sophisticated analysis of the apparent movement of the Moon. By that time, when using the Moon as a reference point for trans-ocean navigation, mariners used sightings of the Moon in conjunction with tabular calculations based on knowledge that the Moon orbits the Earth in 27.3 days, the duration of the sidereal month. The complete orbit of 360 degrees takes 655 hours, giving it an orbital displacement of about half a degree every hour. On this basis the hourly displacement of the Moon is roughly its own diameter in relation to the stars and the Sun. But even a detailed knowledge of the apparent east–west movement of the Moon would be of little use alone in fixing the position of a ship on the open ocean if the ship itself was moving in an easterly or westerly direction. In these circumstances, as a first step, instead of awaiting occasional lunar occultations, early mariners made calculations based on sightings of the Moon in relation to another celestial body. That sighting then had to be compared with the position of the Moon in relation to the same celestial body as it would be observed simultaneously from a known and fixed longitudinal position, which came to be accepted as the Greenwich Meridian. For this purpose, comprehensive predictions of the position of the Moon, in relation to time, as viewed from that meridian, were made available as 'Lunar Tables'. Such Tables were carried and consulted by mariners at all times.

When using Lunar Tables to determine one's position on an otherwise featureless ocean, the first requirement was to measure by sextant the angle between the Moon and, say, a named star. Given the great distance between the observer and the celestial objects in question, this angle, the so-called 'Lunar Distance', is that which is seen from any point on Earth at a particular instant of time, when suitable corrections are made for parallax error. So, entering into the Lunar Tables the 'Lunar Distance' and the apparent altitudes above the horizon of the Moon and the reference star, it was then possible to determine an estimate of the time at the Greenwich Meridian when the same 'Lunar Distance' prevailed. All that was then required was to calculate the time difference between Greenwich Mean Time and local time to estimate the longitudinal displacement of the observer from the Greenwich Meridian. Again, for every hour of the time difference a displacement of 15 degrees of longitude could be assumed. This was the methodology used when mariners had no chronometer allowing them to read off Greenwich Mean Time; they could only estimate that time. They knew only local time, which they were able to validate daily at noon.

It was clear for many years of world ocean exploration that navigation would be improved if, instead of having to estimate the time of day at the Greenwich Meridian, a ship's chronometer, set at Greenwich Mean Time, could provide a reading of that time with reliability. Unfortunately, such a facility was not available to early seafarers for many years. The need for such chronometers was first proposed by Dutch navigators in the mid-sixteenth century, but it was not until John Harrison demonstrated the first sufficiently reliable and prize-winning ship's chronometer in 1759 that ideas became reality. Moreover, it was not until the early part of the nineteenth century that reliable chronometers became routinely available to be used, still with Moon sightings, to determine longitude. By 1798 Samuel Taylor Coleridge epitomized the navigational role of the Moon in *The Rime of the Ancient Mariner* with the words 'she guides him smooth or grim. | See, brother, see! how graciously | She looketh down on him.'

* * *

Given the length of time and degree of sophistication that it took for the human race to establish the ability to navigate by the Moon, perhaps it is not surprising that the notion that some animals might be able to undertake movements orientated to the Moon, let alone navigate by it, is often treated with some scepticism, if not cynicism. Yet homing behaviours that exploit the directionality of Moon-driven tides and orientation movements in direct response to the position of the Moon are increasingly being discovered in even relatively lowly animals. Moreover evidence is accruing that some animals possess biological clockwork of lunar monthly periodicity by means of which they achieve time-compensated homing navigation in relation to the changing position of the Moon.

* * *

Already in an earlier chapter of the book we have seen that sea louse *Eurydice pulchra* swims quite spontaneously in greater numbers just after the times of new and full Moon. It uses its own internal biological clockwork of semilunar periodicity to time its episodes of greatest swimming activity to coincide with the ebbing spring tides. In this way, as the tides recede from springs to neaps the animals are transported down the shore to their preferred homing location in the mid to upper part of a sandy beach. They are not left stranded so high on the beach that they find themselves 'neaped' high and dry in the sand during the lunar quarters. Similarly we have seen how electronically tagged plaice are able to return to their spawning ground by swimming up from the sea-bed when the tide flows in one direction but not in the reverse tidal current. However, there are many other aquatic animals that are at great risk of being swept away from their optimal living locations. In particular, many that reside in estuaries, where river flow exposes them to the risk of being washed seaward, have evolved homing behaviour mechanisms that permit them to stay in or return to their preferred locations. One such behavioural mechanism is that previously described in plaice returning to their spawning grounds, known as 'selective tidal-stream transport'. This is a process that enables a number of estuarine animals,[27] including several of commercial importance, to undertake orientated movements in a homing direction, timed to particular phases of the tidal cycle. Such homing behaviour is, therefore, indirectly related to the lunar cycle. It occurs either in response to a change in environmental conditions as the tide changes, or under the control of the workings of the animal's own biological clockwork that is itself phased to Moon-driven tides.

A classical example of the phenomenon is illustrated by the behaviour of the flounder *Platichthys flesus*, a commercially exploited species of fish commonly found around the coasts of the north-east Atlantic. Populations of this fish have been studied for many years by Dutch fisheries biologists, who, early in the 1900s, recognized that though adults and larvae were to be found in the open sea, juvenile flounders were found in nursery grounds within estuaries of the Dutch rivers, in fresher waters than those of the open sea. Later, biologists set out to discover the mechanism by which this separation of habitats was maintained during the life-cycle of the fish. Correspondingly they sought to ascertain how juveniles homed to their nursery grounds after their initial larval life in the sea. Based upon knowledge that after being spawned in the sea and undergoing early development there as free-swimming larvae, juvenile flounder suddenly appear in large numbers on the wide mud-flats in the Dutch estuaries, extensive observations were made of the numbers collected at various times during the ebb and flow of tides across the mud-flats.[95] It soon became apparent that the numbers of flounder larvae settling out from the plankton as juveniles varied according to the stage of the tidal cycle. Passive transport inshore by tidal action across the mud-flats was excluded because in such circumstances it would be expected that though flood tides would transport larvae inshore, ebbing tides would carry the settlers out to sea again. An active process seemed to be involved and it was found that flounder larvae varied their depth in the water above the mud-flats, depending on the state of the tide. During flood tides they were found near the sea surface and were carried into the estuary by tidal currents, but during the ebbs they were most abundant near the sea-bed, where frictional effects between moving water and the sea-bed ensured that they were able to maintain station in sluggish offshore flows of water. The net effect therefore was that as larvae metamorphosed into juvenile flounders, they were transported together into the estuary. The newly settled juveniles homed effectively in this way, moving in a unidirectional manner towards their nursery grounds. Only after a period of exploitation of the rich feeding grounds of the estuaries do the maturing juveniles migrate seawards again to the adult feeding and spawning grounds where the life-cycle begins again.

Two other commercially important species, in this case estuary residents, that exhibit tide-related homing behaviour against the seaward flow of water in estuaries are the oyster (*Crassostraea virginica*) and the blue crab (*Callinectes sapidus*) on the east coast of the USA. The estuaries in question are influenced by relatively weak tides which induce pulses of dense seawater in and out of the estuaries nearer the bottom, below fresher river water with net flow seawards near the surface. In these circumstances, scientific questions were raised, based on fisheries management considerations, first as to how oyster beds in the Chesapeake Bay area were naturally maintained.

Oysters reproduce by releasing free-swimming larvae into the freely moving water above the oyster beds, which might be expected to transport the larvae seawards. It certainly would be disastrous for the survival of the oyster beds if larvae in large numbers found themselves in the near-surface waters dominated by the outflow of river water: many larvae would be washed out to sea, never to return. How such a disastrous state of affairs is avoided was the subject of a major research project, carried out during the oyster spawning season, from a series of research vessels stationed along the estuary from the oyster beds to the sea. In a co-ordinated programme of sampling, scientists aboard each vessel, working through a series of tidal cycles, pumped water from known depths and for known periods of time, passing the water through filters as they did so. The catches of oyster larvae made by each filter were then kept separate for subsequent analysis.[96] In the event it was found that the oyster larvae spent much of their time in saltier water at lower levels in the water column, at all times avoiding the seaward-flowing fresh water near the surface. During ebbing tides they occurred in greatest numbers in the bottom few metres, where frictional effects of the estuary bed on the flow of water reduced the risks of them being transported seawards. In contrast, during flooding tides, large numbers of larvae were found higher in the water column, where they were gently transported into the estuary by the pulse of sea water pushed in by the flooding tide. Consequently, the net direction of transport of the larvae was towards the parental oyster beds, where they could settle, replenishing the commercial stocks. The question remained, however, as to whether the process of retention in the estuary was entirely passive, the larvae being swirled up into the water column by action of the rising tide, or whether the larvae were actively involved in the process. Certainly, larvae were not swirled up by tidal action during the falling tide, suggesting that they may indeed distinguish between the tidal flood and the tidal ebb, and swim up only during the flood. But was there a way to find out if the larvae were behaving as active 'particles' and not as inactive particles in the water mass of the estuary?

Fortunately it was possible to find an answer to this question since there were known to be truly inactive particles in the Chesapeake Bay Estuary in the form of coal particles lingering from recent historical times when coal was transported extensively along the river. In fact, coal particles were found in some abundance in the samples of larvae that were also counted during the study. Counting of coal particles in samples taken throughout the tidal cycle made the study even more labour intensive but it revealed that inert coal particles and living oyster larvae behaved differently. Coal particles were indeed swirled up from the estuary bed during both the flood and ebb tides, in contrast to oyster larvae which remained near the estuary bed during the ebbing tides. The larvae clearly behaved as living clocks, swimming up from the bottom only during rising tides, either in response to increasing salinity at that time or even under the control of their own internal biological clockwork. Here then is an example of homing behaviour that is related indirectly to the Moon, through the effect of tides. Hence in this case the behaviour of the larvae permits a population of estuarine animals to maintain its integrity over time.

Blue crabs also manage to maintain viable populations in the Chesapeake Bay region, and these again release larvae that swim in the water column. Here, though, the problems of population stability are different from those confronted by the oysters. Whilst adult blue crabs are able to thrive in estuarine waters, their larvae must undergo their early development in fully saline waters of the open sea. Accordingly, egg-bearing females that are about to hatch their young larvae migrate down the estuary towards the sea, where the larvae are released before the adults move back into the estuary again.[97] Once released from the parent females the larvae are dispersed seawards where they undergo several weeks of growth and development through a number of zoea stages, before eventually being swept back towards the estuary mouths by bottom currents induced by tides and winds. There the sea-going zoea larvae moult into the megalopa stage, an intermediate form which begins to acquire the adult crab body shape but which retains the swimming abilities of the zoea larva. At this stage the question arises: how do megalopa larvae find their way into the estuaries to settle on the bottom as juvenile crabs in the vicinity of the adult population? They must do so against the net flow of water down-estuary, which would tend to sweep them out to sea again. Again it seems that they do so by swimming upwards into the pulses of sea water forced into the estuaries by rising tides, remaining in sluggish water near the bottom when the tide ebbs. In this way they undertake homing towards the parental crab stocks where metamorphosis to the first crab stage and stock replenishment takes place.[98] This is another example of homing behaviour indirectly related to the Moon, through the effects of Moon-driven tides.

* * *

Now that there is unequivocal evidence of orientated movements of animals determined indirectly by the Moon, through the influence of tides, is there also evidence of orientated movements by animals in *direct* response to the Moon? There is anecdotal evidence that nocturnal flights by birds and some moths may be orientated by the Moon and the use of stellar cues by night-migrating birds has been well documented for some years.[93] However, until relatively recently very little has been reported in the scientific literature concerning even simple orientation behaviour, let alone navigational behaviour, by animals in relation to the Moon. Possibly the first confirmed example of the phenomenon was not reported until the twenty-first century when the well-regarded scientific journals *Nature* and *Proceedings of the Royal Society* published articles describing the unlikely discovery that the African dung beetle (*Scarabaeus zambesianus*) may have the ability to make directional movements that are orientated by moonlight.[99, 100] Dung, or scarab beetles, are well known to find a pile of dung, say from an elephant, then accumulate a portion of that presumably appetizing food material into a ball, which the insect then rolls away from the dung pile (Fig. 14).

Fig. 14. African dung or scarab beetle rolling a food-ball to safety, navigating in a straight line determined by the position of the Moon.

The advantage to the beetle of moving the ball away from the dung pile is thought to be to provide it with food for later consumption, away from intense competition at the dung pile itself, which is used as food by many insects and other animals. It was when Swedish and South African biologists, led by Marie Dacke of Lund University, carefully observed the behaviour of food-ball rolling by dung beetles that the possibility of orientation by moonlight came to be considered. It was noted that the beetles moved most effectively in straight lines away from the dung pile only on moonlit nights. In contrast, on overcast nights the beetles moved randomly and less far from the dung pile, no doubt with greater risk of loss of the food ball for their own consumption. It therefore seemed that the Moon in some way influenced the most favourable lines of travel of the beetles and the question was—how?

At first sight it might have been thought that the beetles were simply seeing better in bright moonlight, using direct sight of the Moon to determine their direction of travel, however unusual that might seem. However, by careful experiments, it was established, even more surprisingly, that correct orientation was not achieved simply by observing the Moon. By hiding a sight of the Moon from a beetle's field of view or artificially changing the apparent position of the Moon by mirrors, the biologists found that lines of travel away from the dung piles during moonlit nights were unaffected. The beetles were clearly not orientating their movements directly to the Moon's position in the sky. How was this achieved? It is known that some insects, such as bees, are able to recognize polarized patterns of sunlight in the sky, so the question was asked whether dung beetles have eyes that can detect patterns of polarized moonlight, permitting them to orientate accordingly.

Polarization of celestial light occurs when it is scattered by particles in the upper atmosphere. In this way, an ordinary light beam that consists of billions of wave trains of light, all vibrating in different directions, is transformed into a beam in which only those wave trains vibrating in a particular plane are allowed to continue in a now less bright beam of celestial light. So, moonlight (or sunlight) that arrives at the Earth's surface is partially polarized. Moreover, the extent of the polarization viewed from any point on Earth depends on the position of the Moon (or Sun), so there is a pattern of polarization of the sky for any particular time of night or day. To test whether dung beetles were in fact responding to the pattern of polarized moonlight at night the researchers placed above ball-rolling beetles a polarizing filter through which the beetles would see a modified pattern of polarized celestial light. When such a filter was placed so as to rotate the normal sky pattern of polarization by 90 degrees the beetles immediately made right-angled turns in their direction of travel. This behaviour strongly confirmed the notion that the beetles were orientating in response to the night-sky pattern of polarized moonlight. The findings were further backed up by the discovery in the eyes of the beetles of special visual receptors that are able to detect polarized light.

Reports of this discovery in the early twenty-first century claimed, perhaps not surprisingly, that though a number of animals were known to orientate by polarized sunlight, the dung beetle was the only animal so far known that was able to orientate using the million times dimmer polarized light pattern of the moonlit sky.

Some may question the value of such research, especially on the lowly dung beetle, though probably not if the subjects of the study had been pet cats or dogs. Fortunately, though, the importance of such curiosity-driven research remains alive and well, and was occurring elsewhere in the scientific world, even as dung beetles were being studied in Africa. On beaches in

Italy small crustaceans, *Talitrus saltator*, commonly referred to as beach hoppers or sandhoppers, spend their nights foraging among sand dunes above sea level, only to return at dawn to a narrow zone of sand just above sea level, where they burrow. Typically, in the Mediterranean localities where these hoppers occur, the rise and fall of tides is minimal, and it is just above the upper level reached by the tide, in a zone of moist sand, that the hoppers remain burrowed throughout the day. When undergoing their 'zonal recovery' or homing behaviour the hoppers are well known to head directly seawards, and it has been known for some time that they are able to do this in the early morning by orientating to the position of the Sun.[27] However, it is also known that some hoppers return to their burrowing zone on the beach in anticipation of the time of dawn, which raises the question of whether on such occasions they are able to orientate by the Moon. In fact it has been suggested for many years, but not confirmed, that beach hoppers both in the Mediterranean and on Californian beaches in the USA may be able to use lunar direction-finding at night.[101,102] Early experiments showed that hoppers were able to maintain the correct homing direction, even when tested in an experimental arena in the laboratory, where they were denied a view of their normal surrounding coastline but could see the Sun by day or the Moon at night. Critically, hoppers retained the ability to exhibit homing behaviour in the 'correct' direction after several days in constant conditions in the laboratory, at all times being able to compensate for the apparent position of the Sun and the Moon. The implication, even in these early experiments, was that sandhoppers possessed not only a circadian (approximately 24-hour) biological clock that permitted them to compensate for the apparent movement of the Sun, but also a lunar-day (approximately 24.8-hour) biological clock permitting them to compensate for the apparent movement of the Moon. Confirmatory evidence for circadian clockwork in sandhoppers was soon forthcoming[27] but suggestions that the hoppers possessed Moon-related biological clockwork remained speculative for a little longer. However, here is an example in the animal kingdom of how evolution may have pre-empted John Harrison's invention of reliable ship's chronometers that came to the aid of early human navigators.

Direction-finding by animals was first discovered early in the twentieth century but it was not until the mid-twentieth century that true navigational ability by the Sun came to be recognized, initially in birds and later in other animals.[91] Such ability requires an animal not only to orientate to the position of the Sun but to compensate for the apparent movement of the Sun by the use of an internal circadian clock. Acceptance by scientists of this ability came fairly readily, in view of the recognition that circadian clockwork is a common feature of plant, animal, and human physiology. However, acceptance of time-compensated navigation by animals in relation to the Moon was slow, until further critical experiments were carried out early in the twenty-first century by a group of Italian researchers led by Alberto Ugolini at the University of Florence.[103] To study sandhopper behaviour in carefully controlled conditions, hoppers were released at the centre of a level arena that was covered with a large Plexiglass dome to create an artificial sky in the orb of which an artificial Sun or Moon could be illuminated at different positions and intensities. Having previously confirmed that hoppers move in a seawards direction after dawn in relation to the position of the Sun using a circadian biological clock mechanism that allows them to compensate for the apparent positional changes of the Sun, experiments were carried out to establish how seawards hopping occurred before sunrise. Were the hoppers able to navigate by seeing the Moon, compensating for its apparent movement in the sky by possessing not only a solar-day biological clock but a lunar-day clock too?

When tested under the artificial sky at night, hoppers responded to the artificial Moon exactly as if it was the real Moon, hopping towards the edge of the arena on a heading at an appropriate angle to the sight line of the Moon in a direction which would take them seawards if they were above the high-water mark on the beach (Fig. 15).

Importantly, too, the hoppers changed their escape direction towards the edge of the experimental arena when the experimenters made their artificial Moon move on different tracks. Also, since the correct escape direction was maintained during changing positions of the Moon throughout the night,

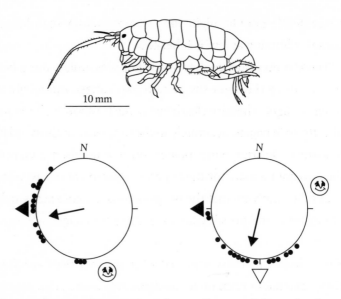

Fig. 15. Laboratory demonstration that sandhoppers (*Talitrus saltator*) navigate seawards by the position of the Moon. The left figure demonstrates the natural situation whereby hoppers in an arena (dots) move in a direction that would take them seawards (arrow) in response to an artificial Moon in its correct azimuthal position. The right figure indicates a 90-degree shift in the orientational direction when the position of the artificial Moon is moved through 90 degrees.

it has to be assumed that the hoppers achieve this ability by possessing another biological clock, additional to that of the solar-day periodicity. They must possess biological clockwork of approximately lunar-day periodicity to enable them to compensate for the apparent movement of the Moon through its nightly arc. Significantly too, it was discovered that hoppers that were born in the laboratory, without experience of life on the beach, were also able to adjust a 'homing direction' under an artificial moonlit night sky, demonstrating that Moon-compass homing ability is inherited.

There is no evidence that beach hoppers, unlike dung beetles, are using the night-sky pattern of polarized moonlight to orient at night. They are clearly using the position of the Moon directly, to 'navigate' accurately in the true sense of the word by following a bearing when possession of a time-sense permits them to compensate for the apparent movement of the Moon over time, implying that the hoppers possess biological clocks of lunar-day and possibly lunar-month periodicity.[128]

* * *

Throughout this book repeated reference has been made to circumstantial or even direct evidence that some animals may possess internal biological clocks of approximately tidal and lunar periodicities. The question now arises as to whether such biological clocks can be shown to have a molecular basis. Is there evidence that tidal and lunar biological clocks have cellular mechanisms similar to that of the daily biological clockwork that drives the circadian rhythms of all living organisms, including ourselves, well described by Russell Foster and Leon Kreitzman in their important book titled *Rhythms of Life?*[35] As hinted in earlier chapters, progress was made on this question concerning lunar and tidal rhythms when in 2013 two critically evaluated papers were published, which demonstrated for the first time that each of these two types of biological clock has a molecular, that is genetic, basis. Lunar molecular clockwork in the marine worm *Platynereis dumerilii* was demonstrated by

Austrian researchers led by Kristin Teissmar-Raible at the University of Vienna. This is one of the 'dancing worms' referred to in Chapter 4 that performs its nuptial spawning dances at the times of new Moon.[31] The circalunar (monthly) clock in *Platynereis* was shown to be quite distinct from the circadian clock of the worm, but modulated its timing.

The nature of circatidal rhythms has been long debated, some arguing that they might be controlled by circadian clocks, not by circatidal clocks at all. It was suggested that two circadian clocks of approximately 24.8-hour periodicity, operating in antiphase, could explain the apparent expression of circatidal rhythms of 12.4-hour timing.[29] Others favour a dedicated circatidal clock operating in addition to a circadian clock which times the day-to-day activities of an animal.[27] A solution to this dilemma awaited input from clever scientists who were able to devise a method of stopping the circadian clockwork of an animal and observing whether its tidal rhythm also stopped or continued running under the control of separate circatidal clockwork. A group of such British biologists came together to address the problem and, at the outset, one member of the team, Charalambos Kyriacou of the University of Leicester, making a wager with another member, Michael Hastings of Cambridge University, that they would find no true tidal clockwork. These two scientists, with others, investigated the problem in the sea louse *Eurydice*, referred to in Chapter 3, which exhibits tidally related behaviour to determine its time of burrowing during the falling tide and which also has a circadian rhythm that controls its day/night-related patterns of colour change. Using techniques of modern genetics they established that the two types of rhythm in the sea louse were each controlled by molecular clockwork and that the circadian clock controlling colour change could be disabled without affecting the circatidal clock.[30] So Kyriacou lost his bet; *Eurydice* does have true tidal clockwork. Moreover, recent discoveries by the same group of researchers have provided the first molecular leads for dissecting the tidal clockwork of *Eurydice* and beginning its characterization.[125]

So, just as it has been established for some years that there are specific clock genes controlling the basis of circadian behaviour in living organisms, including Man, it can now be concluded that there are also designated clock genes that control molecular mechanisms forming the basis of tidal and Moon-related behaviour discussed throughout this book. Hence, while it has been clear for some time that living organisms have adapted throughout evolution to daily cycles induced by rotation of the Earth in relation to the Sun, it now seems fairly conclusively demonstrated that genetic adaptation has also occurred in relation to cycles of the Moon and Moon-generated tides. Furthermore, with the discovery of clock genes controlling tidal and lunar rhythms, albeit so far only in two marine animals, such genes are likely to occur more widely in animals, possibly too in humans.

In the last chapter we shall consider reported accounts of the impact of the Moon on modern human life, seeking to judge whether any supposed impacts can indeed be considered to provide evidence of adaptation to lunar cycles with the incorporation of lunar biological clockwork into the human genome.

9

THE MOON AND THE HUMAN CONDITION

With a long history of perceived associations between lunar cycles and the living world, characterized in legend and folklore, it is not surprising that beliefs in such linkages persist among some people even today, not least in some modern assertions that the Moon has influence over aspects of human life. Equally understandably, sceptics have long challenged the likelihood that the Moon influences the human condition. Not only lay people, but some scientists too have commented that there is no solid evidence that human biology is in any way influenced by the lunar cycle,[4,104] and questions are often raised as to how the Moon could possibly influence life processes. As we have seen, however, scientific experiments have begun to demonstrate convincingly that the Moon does influence the behaviour of some animals, both directly and indirectly. With such evidence, in the twenty-first century some scientists and medical practitioners are again beginning to consider the possibility that there may indeed also be causal relationships between the Moon and some aspects of human biology and medicine. Let us begin by considering the perpetually recurring assertions that the human menstruation cycle has associations with the cycle of the Moon.

From the age of puberty until the menopause, one half of the human population on Earth experiences a 'monthly' pattern of menstruation and fertility. It is again not surprising, therefore, that questions are asked afresh about the possible relationship between that normally regular female experience and the monthly phases of the Moon. As discussed in Chapter 1, Aristotle posed the question but was able to conclude, unequivocally, that if the time of menstruation and the waning of the Moon coincided, that was by chance, since menstruation could occur at any other phase of the Moon. Yet the question has been and continues to be raised. In his writings, as briefly mentioned in Chapter 1, Charles Darwin asked, if humans are descended from fish-like ancestors, as his theory of evolution proposed, then why should the feminine cycle not be a vestige of the past when life depended on the tides and therefore on the Moon? In fact, the length of the human menstrual cycle is variable and its often quoted average duration of 27.32 days is a closer match to the duration of the sidereal month (27.3 days), the time taken for the Moon in its orbit to return to the same position in space above the Earth, than to that of the synodic or true month (29.5 days). The tides, of course, do not follow the sidereal month. They follow the synodic or true month, which takes account of the fact that the Earth and the Moon move together in their orbit around the Sun, ensuring that the Moon does not regain its original position above an observer on Earth until two days beyond the sidereal month. The average duration of the human menstrual cycle is therefore shorter than the perceived lunar month from full Moon to full Moon, which is also reflected in the monthly tidal cycle. Hence, over time, average human menstrual cycles will constantly change their phase in relation to the true lunar cycle. Correspondingly, too, individual cycles of slightly longer than average periodicity may show correlations with the lunar cycle, but others will not. Nevertheless there have been repeated searches for correlations.

For example, a 1986 study of 826 women aged between 16 and 25 years with 'normal' menstrual cycles, resident in Beijing and Guangchow, China, reported that 28.3 per cent menstruated around the time of new Moon, a percentage stated to be 'significant'.[105] In addition, an earlier study of 312 women in the USA, women who were selected from a sample because their menstrual cycle duration approximated to the 29.5 day periodicity of the true month, 'tended' to menstruate during the dark phase of the Moon from the last quarter, through new Moon, to the first quarter.[106] In contrast, a 2001 review of the subject concludes unequivocally that 'in apes (including humans) no temporal relationship emerges between the duration of the lunar month in the present geologic era and the menses'.[107] Evidence to support this conclusion was obtained by comparing the length of the menstruation cycle in humans with that of other higher animals to which *Homo sapiens* is evolutionarily related.

The outcome of such studies is that the menstrual cycles of other primates are highly variable between members of the group, ranging from 10 to 50 days. The three types of ape closest to humans in evolutionary terms are orang-utans, which diverged from the common ancestor 11–14 million years ago; gorillas, which diverged about 8 million years ago; and chimpanzees, which diverged 6–7 million years ago or more. We cannot know the length of the menstrual cycles of extinct human relatives such as *Homo erectus* and *H. neanderthalensis*. However, it is known from observations of individual animals living in zoos that the average lengths of the cycles of the next nearest human relatives are: orang-utans 29–32 days, gorillas 43–49 days, and chimpanzees 31–37 days, compared with the average of 27–28 days, with great variability, in humans. None of the menstrual cycles reported for primates, including humans, can be considered to coincide with the true lunar month and all vary so considerably that it seems unlikely that any are controlled by, or 'phase locked' to, the lunar cycle.[106] Indeed, among many chronobiologists, a problem arises concerning data collection, interpretation, and statistical analysis of postulated lunar relationships of human life processes. Modern studies of animal and plant biological rhythms are usually based on long time-series of data to which rigorous statistical procedures can be applied whereas studies of purported Moon-related rhythms in humans are often demographic, with sometimes questionable statistical outcomes.[48] There is, therefore, justified scepticism among many scientists concerning supposed relationships between human life processes such as menstruation and lunar cycles.

The most economical conclusion to be drawn is, in fact, that of Aristotle. Because of the superficial similarity between the human menstrual cycle and that of the Moon, when a correlation occurs it is by chance and not because there is a causal relationship between the two. The words 'menses' and 'menstruation' merely perpetuate beliefs that the occurrences are Moon-driven, inevitably prompting further investigations, even today.

Similarly, there is no consistency in recent studies of the timing of natural childbirths, which, in historical times, when the Moon was called the 'great midwife', were said to occur mainly around times of the full Moon. On the one hand, a 1998 article by obstetricians in Italy claimed to support that folklore belief when they reportedly found a connection between the distribution of spontaneous full-term deliveries and the lunar month.[108] In a sample of 460 women who had each borne several previous children, the days of delivery at childbirth were found to occur around the first or second day after the time of the full Moon. The records were said to be statistically significant but were nevertheless too weak to permit predictions of birth dates to be made in individual cases. However, other studies in Italy around the same time gave conflicting

results.[4] Notably, in a much larger sample of 7,842 spontaneous deliveries at an Obstetric and Gynaecology Clinic of the
University of Florence no significant differences were found
between the days of birth and the days of the lunar cycle.[109]
Early beliefs that the lunar time of conception could be used to
predict the gender of the child are also without justification.

It is clear that ancient myths relating to the Moon and
human reproductive processes are deeply engrained in the
human psyche. Equally, where many such beliefs have been
considered rationally they have been rejected as lacking in
scientific justification. As a result, scepticism has built up
concerning any linkages at all between the Moon and human
biology. Nevertheless, human curiosity, not least among some
medical practitioners and scientists, is such that investigations
of possible linkages to human health continue. During the
latter half of the twentieth century, for example, analyses
were made of extensive records of the times of death of people
from Romania and the Czech Republic, to ascertain whether
there was any relationship between the time of death and the
lunar phase.[110] In the Romanian study, medically recorded
times of death of about 1.8 million people during the period
1989–95 were analysed. Apart from a clear annual cycle of
deaths, with greatest general mortality occurring during winter
months, several other cyclical patterns were observed, notably
a twice-monthly increase in mortality, peaking around times of
the first and third quarters of the Moon. At best, the authors
concluded only that the semilunar maxima 'appeared to be

significant', but pointed out that the findings agreed closely with those from a similar study in the Czech Republic. In the latter study, 1,437 sudden deaths from cardiovascular mortality in the city of Brno were recorded over the period 1975–83.[111] To standardize the data, the study included only sudden deaths due to cardiovascular disease when people died at home or elsewhere but not in hospital. In other words the deaths were due to coronary disease before there was any medical intervention. Here, again, there was a twice-monthly pattern of mortality, with greatest numbers occurring just before times of the first and third quarters of the Moon, and fewest just before new and full Moon.

Such cycles in the Romanian and Czech data are significant only after sophisticated statistical analysis and have yet to be substantiated by comparable studies elsewhere. Observations such as these have nevertheless prompted additional clinical investigations reported in early twenty-first-century medical cardiovascular literature. One such line of investigation concerns possible lunar variations in the timing of ruptures of aneurysms, asking whether the phenomenon is 'myth or reality'. An aneurysm occurs when a bulge appears at a weak position in the wall of an artery, particularly the aorta, spontaneous rupture of which has potentially serious consequences for the patient. Yet again, though, the findings of such studies are equivocal. Some investigators in 2008 reviewed retrospectively the medical records of 111 patients admitted to a neurosurgical department over a period of five years because of aneurysmal subarachnoid haemorrhage. Peak incidence of aneurysm rupture (28 patients) was seen during the new Moon phase, a number which was reportedly statistically significant.[112] In contrast, a year later another group contradicted that conclusion, stating unequivocally that 'the impact of the lunar cycle on aneurysmal subarachnoid haemorrhage is a myth rather than reality'.[113] Nevertheless, curiosity about possible links between cardiovascular disease and the lunar cycle has continued unabated. For example, a team of surgeons in the USA in 2013 undertook a study with a seemingly esoteric objective, based on an initial assertion that 'the effect of the lunar cycle . . . on ascending aortic dissection surgery outcomes is unknown'.[114] Their results, based on medical records from Rhode Island Hospital, purported to show that patients having aortic surgery, with or without associated aortic valve repair and coronary bypass surgery, had increased chances of survival and reduced lengths of stay in recovery if their operations were carried out during or just after a full Moon stage of the lunar month. As in earlier reports from Eastern Europe, however, the underlying mechanisms of differential survival rates associated with the lunar cycle 'remain elusive'.

A further common belief that has been deeply ingrained in human thinking over historical time is the supposed linkage between madness and the Moon, as evidenced by use of the word 'lunatic' derived from the Latin word for the Moon, 'luna'. In Greek society Hippocrates considered that moonlight caused nightmares, perpetuating a myth that it was dangerous to sleep outside under the light of the full Moon. Then in the sixteenth century the human brain came to be called the 'microcosmic Moon'. It was clearly thought that human behavioural changes were under the influence of the Moon and that madness would be intensified during the full Moon. However, reason was beginning to be brought to bear on the matter early in the seventeenth century, when, for example, William Shakespeare in *King Lear* gave Edmund the lines:

> This is the excellent foppery of the world, that, when we are sick in fortune, often the surfeit of our own behaviour, we make guilty of our own disasters the sun, the moon and the stars.

Even so, the connection between human behaviour and the Moon was readily accepted until as late as the end of the eighteenth century, so that, in occasional criminal cases, exceptions from punishment were allowed by some legal systems if the crime was committed during a full Moon period. The belief was so entrenched that keepers of early lunatic asylums adopted it too, to the extent that they considered it necessary to employ additional asylum staff during times of the full Moon to ensure that inmates were kept under control.[4] Certainly it is not inconceivable that prior to modern urban lighting the Moon was a significant source of nocturnal illumination that caused sleeplessness and disturbed behaviour during times of full Moon.

Indeed, some convincing modern research addresses this very problem and claims to provide the first evidence that the monthly lunar cycle can modulate the pattern of sleep in humans.[115] The researchers start from the widely accepted view that there is a great deal of folklore but no consistent association of Moon cycles with human physiology and behaviour. They then claim to show that subjective measures of sleep, by observation of volunteers, and objective measures, recorded by electroencephalogram (EEG), both vary with lunar cycles. The authors claim that their findings present the first reliable evidence that a lunar rhythm can modulate sleep structure in humans, with the implication that sleep patterns reflect the possession by humans of inherent biological clockwork of approximately lunar month (circalunar) rhythmicity. The observations were made under the extremely rigorously controlled constant conditions of a modern chronobiological laboratory, that is without time cues and without investigators or sleep volunteers being aware of the analyses to be made in relation to lunar phase. It was found that around the times of full Moon EEG activity during deep sleep decreased by 30 per cent, time taken to fall asleep increased by 5 minutes, and total sleep duration, assessed by EEG, was reduced by 20 minutes. This discovery (Fig. 16) of apparently heritable Moon-related clockwork in humans was supported by a later study, which showed that sleep time around full Moon decreased spontaneously by 25 minutes and cortical reactivity to environmental stimuli during sleep increased, though with some qualifications related to individual susceptibility.[116]

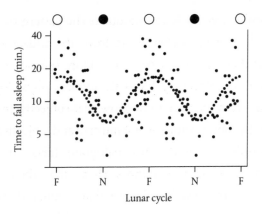

Fig. 16. Human volunteers in a sleep laboratory, in the absence of time cues, fall asleep more quickly during nights of new Moon.

Inevitably, though, these recent discoveries were challenged by sceptics unable to repeat the findings, who add suggestions that bias is introduced into the literature by selective publication of data.[117] It is stated by these authors that in such studies there is a tendency to publish positive, but not negative or non-confirmatory data: the so-called 'file-drawer problem'. Countervailing this statement, however, it has to be said that there are reports of scientists being reluctant to publish positive data concerning Moon-related biological phenomena in fear of being regarded as members of the 'lunatic fringe'. Clearly, the debate continues and must take account of the fact that there are as yet no known mechanisms whereby, say, bright moonlight at full Moon synchronizes the circalunar rhythm. If the findings of Moon-inhibited sleep are substantiated by future research, they may contribute significantly to our understanding of Moon-related health phenomena in humans, suggesting perhaps that some of the myths may indeed have to be considered as real.

* * *

Based not on myth but on sheer practicality, even as late as the early nineteenth century in Britain the influence of moonlight was apparent on significant aspects of human behaviour. From 1776 to 1820 members of the so-called Lunar Society of Birmingham met regularly on the Monday nearest to the time of full Moon.[118] The members were in no sense Moon-worshippers but were eminent scientists, engineers, and industrialists of the day who met to enjoy mutual intellectual stimulus as they sought to advance their particular scientific, engineering, or entrepreneurial projects. Their meetings were pragmatically timed to the evenings of the full Moon to ensure that there was sufficient moonlight afterwards to avoid the risk of attack by vagabonds as they made their way home. Members of the Society were highly influential during the Industrial Revolution in Britain during the period from 1760 to 1840. 'Lunaticks' they may have been called, but their endeavours were probably responsible for the subsequent introduction of street lighting in urban Birmingham, which removed the justification for the timings of their meetings in the first place. Thus they no doubt contributed to the development of what has come to be called 'light pollution' in the night skies of towns and cities today. In one sense, therefore, the Lunar Society can be considered to have been at least partially responsible for a reduction in the awareness in many modern urban societies of the changing pattern of moonlight throughout the lunar month, and hence to the modern reduction of beliefs in lunar myths.

Nevertheless, beliefs about the Moon's influence on human behaviour persist in some present-day societies, industrialized or not. As recently as the end of the twentieth century a questionnaire sent to people in New Orleans, USA, concluded that nearly 50 per cent of the sample held the opinion that lunar phenomena do affect human behaviour. Notably, it emerged that mental health professionals, including social workers, clinical psychologists, and nursing aides, all held this belief quite strongly when compared with other societal groups who were not working with psychiatric patients in a professional capacity.[4] Even so, there has been little success in confirming lunar associations with various behavioural phenomena, despite considerable research effort. For example, fluctuations in emergency admissions to a suburban community hospital in the USA, analysed over a four-year period, showed no correlation with times of the full Moon, despite preconceptions to the contrary.[119] Similarly in a retrospective one-year study of over one thousand trauma cases in another US hospital in Pittsburgh, victims of assault by blunt weapons, gunshot wounds, and stabbings were analysed to assess whether there was any correlation between violent human behaviour and the lunar cycle. Again it was found that such preconceptions were statistically unfounded.[120] Also, a nine-year study of car accidents in Canada and another study of drug-overdose cases in the USA showed none of the Moon-based correlations that, anecdotally, have sometimes been attributed to them.[4] In contrast, a British study published in the *British Medical Journal* in 1984 did show a statistically significant increase in crime rate in rural, urban, and industrial localities during full Moon days, but with no explanation as to why that might have occurred.[121]

Today, consistent with the suggestion from recent research that the Moon may influence sleep cycles, and perhaps emanating from older views that the Moon induces madness, a recurring theme when considering Moon-related behaviour concerns possible lunar links with psychiatric diseases and suicide.[4] In fact, contrary to popular belief, several studies have been carried out which show that suicide occurrence does not conform to the lunar cycle. A particularly extensive study of 4,190 suicide deaths during the period 1925–83 clearly showed this to be the case.[122] Similarly, fifty years of research have shown no association between psychiatric illness and the Moon.[4] Nevertheless practitioners of psychiatry report that mental illness is still occasionally attributed by some to the Moon, an association apparently fostered by some hospital staff, patients, and their families, particularly concerning patients diagnosed with schizophrenia. Such beliefs may arise even now, despite lack of evidence in their support, because part of the problem for psychiatrists in managing mental illness and for affected families trying to cope with the illness is that symptoms are often intermittent. As a consequence of this, one of the 'coping mechanisms' for dealing with the illness is to seek a pattern and external cause, such as waxing and waning of the Moon, a linkage which becomes reinforced by repeated assertions by members of the community and by the press.[123] Nevertheless, a connection between the behaviour of mentally ill patients and the Moon, though a long-standing belief, remains a cultural myth and has no scientific standing.[123]

The same is true of the medical condition of epilepsy, which for many years was also considered to be of supernatural origin, under the influence of the Moon. That myth, too, has persisted despite the fact that even Hippocrates considered epilepsy to be a clinical condition due to natural, and not supernatural, causes. Scepticism concerning lunar myths in relation to human medicine originated therefore at least as early as the fifth century BC. Yet books and articles continue to be written postulating that the full Moon may either cause or exacerbate mental illness, particularly in patients diagnosed with schizophrenia, sometimes to the extent that fanciful explanations are put forward to suggest how lunar effects are induced. While it is not inconceivable that moonlight itself might influence some aspects of human behaviour, as earlier chapters have demonstrated in some animals, imaginative suggestions that the Moon has influence over the human body's electrical field, magnetic field, and gravitational field are pure speculation. No evidence is put forward to suggest how weak environmental signals of this kind might be sensed, even if they are universal.

Just as there are unsubstantiated claims that tree stem diameters fluctuate with 'tide',[124,16] there are even suggestions that the Moon induces 'tides' in the human body because of its high fluid content. This is a concept that can be traced back to Greek philosophers, and over time has been used to try to explain the fluctuating incidence of various human diseases and behavioural traits. In fact, the gravitational pulls of the Moon and the Sun are so small that they affect the human body only by the weight equivalent of a bead of sweat or a human hair. Repeated application of even such small forces is able to induce significant tidal resonance in the open ocean, but lunar gravitational pull is insufficient to affect significantly the enclosed fluids of the human body. Nevertheless the sway of mythology often still prevails.

Another popular perception is of a connection between the Moon and human libido. Jack Palmer, a Professor of Chronobiology at the University of Massachusetts, together with a collaborating Social Sciences colleague, carried out a scientific study to test this notion.[29,126] In an experiment published in 1982 they reported their successful investigation during which they paid seventy-eight couples the princely sum of one dollar each to mail in a postcard reporting if and when they had had sexual intercourse during the previous twenty-four hours. They collected 5,584 days of data, in which 1,941 episodes of copulation were reported to have taken place. Perhaps not surprisingly, the greatest number of copulations occurred between 21.00 and 01.00 hours, but it was concluded that there was no significant correlation between the daily or monthly phase of the Moon and human libido, thus shattering another myth concerning the Moon and the human condition.

* * *

In conclusion, beliefs in the influence of the Moon on human life, that have their origins in the increasing perception of the natural world by ancient peoples, have so far largely failed to find validation in scientific studies.[104] This, together with claims based on poorly designed experiments and data analysis,[48] has encouraged modern-day scepticism of such notions. Yet the phenomenon of lunar periodicity in animals has certainly moved out of the 'realms of mysticism' and has generated testable hypotheses, as Carl Hauenschild predicted when he published his pioneering experiments on 'dancing worms' in 1960.[56] In recent decades, as we have seen, it has become acceptable that indirect effects of the Moon, through tides, and direct effects of the Moon, through changing moonlight intensity, are able to induce patterns of behaviour in animals and perhaps plants that match the lunar day and the lunar month.[27] Circumstantial evidence also suggests very strongly that, since such patterns are able to persist in organisms maintained in constant conditions in the laboratory, they are partly controlled by internal biological clockwork and are inherited.[27] Now, in recent years, direct evidence has suggested, for a few marine animals that Moon-driven tidal and lunar monthly rhythms are indeed heritable, controlled internally by specific molecular mechanisms in the form of Moon-related circatidal and circalunar clock genes.[30,31,125] So, if it is confirmed that such Moon-related genes are widespread in marine organisms, there is reason to suppose that they, like circadian clock genes, might be widespread throughout the animal kingdom, including in humans. The mystery and fascination for the subject remains to tantalize laymen and scientists alike, and the search for answers continues, despite widespread scepticism.

But are attitudes changing? As we have noted, very recent investigations of human sleep patterns in rigorously controlled conditions have shown that subjective and objective measures of sleep vary according to the phase of the Moon and may reflect circalunar rhythmicity in humans.[115,116] The findings have been challenged[117] but, if confirmed, they raise the possibility for the first time that humans possess Moon-related (circalunar) clock genes as well as Sun-related (circadian) clock genes. If so, there is clearly much more to discover about how organisms other than those in the sea, including humans, use their biological clocks to regulate their behaviour, and how desynchronization of the clocks may lead to disease. Perhaps there may yet prove to be something behind the folklore concerning human behaviour and the Moon if the periodicity of the Moon, like that of the Sun, is embedded in our genes.

GLOSSARY

aneurysm a bulge in a weak spot of an artery.

antiphase here used in connection with a (so far unsubstantiated) hypothesis explaining how circatidal (approximately 12.4-hour) rhythms are controlled by two circadian (approximately 24-hour) rhythms run 12 hours out of phase with each other such that one expresses peaks when the other expresses troughs.

apogee the point in the elliptical orbit of the Moon when it is farthest from the Earth.

arms race in biology, the evolutionary process whereby, say, a predatory animal species develops adaptations which improve its ability to capture prey, and the prey species develops competing adaptations which improve its ability to avoid predators.

azimuth the compass-bearing to a point on the horizon vertically below a celestial object such as the Moon.

Big Bang the 'singularity' defined by physicists as the origin of the universe, some 4.5 billion years ago.

Big Splash an event thought to have occurred over 4 billion years ago, soon after the earth was formed, during which the proto-earth was impacted by a cosmic object of at least the size of Mars which threw up debris into space which coalesced to form our orbiting Moon.

biodynamic agriculture modern methodology based on folklore purporting to relate plant growth to lunar cycles.

chromatophore cell containing pigment which can expand or contract to change the colour of an animal.

chronobiology the study of biological rhythms.

circadian rhythm/clock a biological rhythm of approximately daily (24-hour) periodicity expressed in constant conditions, under the control of an internal biological clock system of that periodicity.

circalunar rhythm/clock a biological rhythm of approximately lunar (29.5 day) periodicity expressed in constant conditions, under the control of an internal biological clock system of that periodicity.

circasemilunar rhythm/clock a biological rhythm of approximately half-monthly (14.5 day) periodicity expressed in constant conditions, under the control of an internal biological clock system of that periodicity.

circatidal rhythm/clock a biological rhythm of approximately tidal (12.4 hour) periodicity expressed in constant conditions, under the control of an internal biological clock system of that periodicity.

clock gene gene component of the molecular mechanisms that produce circadian, circatidal, and other biological rhythms.

dendrochronology the study of growth rings, for example in trees, clams, and corals, as indicators of past climate.

DVM diurnal vertical migration of plankton in the sea, rising towards the surface by night and sinking by day.

environmental synchronizer a rhythmic environmental variable such as light, temperature, tide, or moonlight, which in nature sets the timing of biological clocks.

grunion a remarkable marine fish which mates and lays its eggs high on sandy beaches at times of highest tides, associated with full and new Moon.

liposome fatty vesicle, artificially constructed, or naturally occurring in living cells.

lunar day the average interval of 24.8 hours between, say, successive moonrises, that is one rotation of the earth in relation to the orbiting Moon.

lunar occultation occasions when, as seen from Earth, the Moon obscures another celestial object such as a planet.

Lunar Society an eighteenth-century society, originating in Birmingham, UK, in which scientists, engineers, and entrepreneurs met during evenings timed to coincide with the full Moon.

Lunar Tables tables carried by early mariners which provided comprehensive time-based predictions of the position of the Moon, in relation, say, to a named star ('lunar distance'), as viewed from the Greenwich Meridian.

megalopa the last stage in the life history of a crustacean such as a crab when, typically, it reverts from swimming in the plankton to settling on the sea-bed.

metonic cycle a nineteen-year cycle which arises because lunar months are shorter than calendar months. Accordingly, after a particular phase of the Moon, such as full Moon, occurs on a particular calendar day, that coincidence is not repeated until nineteen years later.

molecular clock see *clock gene*.

monomer a simple molecule consisting of a single chain, which can exist alone or can combine with other monomers to form a *polymer*.

Moon Illusion the phenomenon whereby the full Moon at the horizon appears larger than when it is higher in the sky.

neap tides the smallest tides of the month, which occur every two weeks, just after times of the lunar quarters.

neuston mostly microscopic animals that live near or on the sea surface.

nuptial dances breeding behaviour of normally sea-bed-living worms which swarm in the plankton, releasing eggs and sperm for fertilization as they do so.

ocean skater a small insect like a pond-skater that is sometimes caught in *neuston* nets on the sea surface.

ocean weather fluctuations in currents and turbulence in the sea, often attributable to monthly tidal cycles.

otoliths calcareous granules in the inner ear of vertebrates which, in conjunction with sensory cells, permit assessment of position in relation to gravity. In fish, after appropriate preparation, otoliths reveal growth rings which can be used to determine age.

perigee the point in the elliptical orbit of the Moon when it is nearest to the Earth.

phase-locked here describing a rhythmic biological phenomenon that coincides with a rhythmic environmental variable.

planetisimal a very small planetary body which theoretically could merge with similar bodies to form a planet.

polarized light when a light beam of many wave-trains, each vibrating in a different direction, is transformed into a beam in which the wave-trains vibrate in the same plane, as when sunlight or moonlight are polarized when passing through the Earth's upper atmosphere.

polymer a large organic molecule consisting of repeated sub-units (*monomers*).

residual periodic variables minor environmental variables such as atmospheric pressure and cosmic ray bombardment that are not kept constant in so-called constant conditions in the laboratory.

sea louse a small marine isopod crustacean, so called because of its close affinity with more familiar terrestrial species of woodlouse.

selective tidal-stream transport unidirectional transport of planktonic animals that swim up into the water column during tidal flow in one direction and sink to the sea-bed, remaining stationary, when the tidal currents flow in the reverse direction.

sidereal month the time taken for the Moon to orbit the Earth, 27.3 days, ignoring the Earth/Moon travel in their orbit of the Sun.

spawning window time during the year when an organism is potentially able to spawn if the correct environmental conditions present themselves.

spring tides the largest tides of the month which occur every two weeks, just after times of full and new Moon.

synodic month the interval between, say, successive full Moons, 29.5 days, which is longer than the *sidereal month* because it takes account of the Earth/Moon travel in their orbit of the Sun.

syzygy the alignment of the Sun, Moon, and Earth in a straight line, the combined gravitational pulls of the Sun and Moon generating spring tides on Earth. In combination with the cycle of the Moon's proximity to Earth, a 'syzygy inequality cycle' (SIC) is generated whereby the largest spring tides of the month alternately coincide with new and full Moon on a fourteen-month cycle.

tidal memory reflects the ability of an organism to express tide-related behaviour when kept away from the influence of tides, under the control of its inbuilt biological clockwork of approximately tidal periodicity.

zoea one of several early stages in the life history of a crustacean such as a crab which typically swims freely and feeds in the plankton before eventually moulting to the *megalopa* stage, when it settles on the sea-bed.

zonal recovery the ability of, say, a beach-living animal to navigate back to its resting zone on a beach after foraging widely above or below that zone.

zooxanthellae yellow or brown single-celled algae living symbiotically in the tissues of certain animals.

NOTES AND REFERENCES

1. Berno, F. R. (2001). Seneca and the moon: the cultural importance of our satellite. *Earth, Moon and Planets*, 85–6, 499–503. Expounds the views of a Roman stoic philosopher of the first century AD who knew of the Moon's movements and related them to tides and eclipses.
2. Duhard, J.-P. (1988). Le calendrier obstetricale de la femme a la Corne de Laussel. *Bulletin de la Société Historique et Archéologique du Périgord*, 65, 23–39. Interpretation of Palaeolithic cave art.
3. Graves, R. (1955). *The Greek myths*, Vols. I and II. Penguin Books. A comprehensive interpretation of Greek mythology, with notable references to lunar deities.
4. Zanchin, G. (2001). Macro and microcosmus: moon influence on the human body. *Earth, Moon and Planets*, 85–6, 453–61. A sceptical, historical review of lunar effects on the human body, with some suggestions for further study.
5. Brueton, D. (1991). *Many moons: the myth and magic, fact and fantasy of our nearest heavenly body*. Prentice Hall Press, New York. Discusses Darwin's reference to the observed similarity between lunar and human menstrual cycles.
6. Carroll, W. C. (2001). Goodly frame, spotty globe: earth and moon in Renaissance literature. *Earth, Moon and Planets*, 85–6, 5–23. Reflects the change in Renaissance literature from pre-Galileian personification of the Moon to demystified conceptions created by the discoveries of Galileo.

7. Romano, G. (2001). The moon in the classic Mayan world. *Earth, Moon and Planets*, 85–6, 557–60. In the Classic Period of Mesoamerica (AD 250–900) the Maya demonstrated particular attention to the movements and appearance of the Moon, which was considered an important goddess.

8. Bellinati, C. (2001). The moon in the 14th century frescoes in Padua. *Earth, Moon and Planets*, 85–6, 45–50. Interpretations of lunar imagery in fourteenth-century frescoes in important buildings in Padua, Italy.

9. Jalufka, D. A. and Koeberl, C. (2001). Moonstruck: How realistic is the moon depicted in classic science fiction films? *Earth, Moon and Planets*, 85–6, 179–200. Representations of the Moon in classic films of the science fiction genre up to and including Kubrick's *2001: A Space Odyssey*.

10. Montgomery, S. L. (1994). The first naturalistic drawings of the moon: Jan van Eyck and the art of observation. *Journal for the History of Astronomy*, 25, 317–32. Representations of early naturalistic sketches of the Moon based on critical observation.

11. Olson, R. J. M. and Pasachoff, J. M. (2001). Moonstruck: artists rediscover nature and observe. *Earth, Moon and Planets*, 85–6, 303–41. Discussion of early artistic depictions of the Moon based on critical observation rather than relying on stylized representation.

12. Pigatto, L. and Zanini, V. (2001). Lunar maps of the 17th and 18th centuries. Tobias Mayer's map and its 19th century edition. *Earth, Moon and Planets*, 85–6, 365–77. Early lunar cartography following Galileo's initial observations by telescope.

13. Agel, J. (ed.) (1970). *The making of Kubrick's 2001*. Signet: The New American Library, New York. Depictions of realistic topography incorporated into lunar imagery in movie-making.

14. Wilhelms, D. E. (1993). *To a rocky moon—a geologist's history of lunar exploration*. University of Arizona Press, Tucson. A salutary tale from the 'space race' during the Cold War between the USSR and the West.

15. Speiss, H. (1990). Chronobiological investigations of crops grown under Biodynamic Management. I. Experiments with seeding dates

to ascertain the effects of lunar rhythms on the growth of Winter Rye. *Biological Agriculture and Horticulture*, 7, 165–78. Article purportedly supporting the notion of 'Biodynamics', which recommends Moon-phased plantings of crops, but rejecting the proposition that planting success varies with the position of the Moon in relation to star constellations.

16. Zurcher, E., Cantiani, M.-G., Sorbetti Guerri, F. and Michel, D. (1998). Tree stem diameters fluctuate with tide. *Nature*, 392, 665–6. A review of the chronobiology of trees, including claims that aspects of tree growth vary in relation to lunar phase.

17. Hawking, S. (1988). *A brief history of time*. Bantam Press. Late twentieth-century thinking on cosmology, with Galileo's early seventeenth-century contribution set in context.

18. Longo, O. (2001). Ancient moons. *Earth, Moon and Planets*, 85–6, 237–43. Ancient views of the nature of the Moon by philosophers and scientists from classical antiquity.

19. Taylor, S. R. (1992). *Solar system evolution*. Cambridge University Press. Develops a consensus argument favouring the 'Big Splash' origin of the Moon.

20. Lissauer, J. T. (1997). It's not easy to make the moon. *Nature*, 389, 327–8. Expresses reservations about the 'Big Splash' hypothesis.

21. Benn, C. R. (2001). The moon and the origin of life. *Earth, Moon and Planets*, 85–6, 61–6. A short review favouring the 'Big Splash' origin of the Moon and the origin of life in intertidal pools influenced by Moon-driven tides.

22. Lovelock, J. (1988). *The ages of Gaia*. Oxford University Press. Lovelock's continuing development of the Gaia Hypothesis, centred on the proposition that plants, animals, and their environments are all part of one 'organism'.

23. Oro, J. and Lazcano-Araujo, A. (1981). Cometary material and the origins of life on earth. In: C. Ponnamperuma (ed.), *Comets and the origin of life* (D. Reidel, Dordrecht), 191–225. Speculation concerning cosmic sources of the water that aggregated on Earth as it cooled.

24. Chyba, C. F. (1990). Impact delivery and erosion of planetary oceans in the early inner Solar System. *Nature*, 343, 129–33.

Discusses the implications of the closer proximity of the Moon to the Earth than now, at the time of the origin of life.

25. Oro, J., Miller, S. L. and Lazcano, A. (1990). *Annual Reviews of Earth and Planet Science*, 18, 317. Review of ideas and laboratory studies concerning possible origins of self-replicating living systems from 'primordial soup'.

26. Deamer, D. W. and Barchfield, G. L. (1982). Encapsulation of macromolecules by lipid vesicles under simulated prebiotic conditions. *Journal of Molecular Evolution*, 18, 203–6. Reports on encapsulation of DNA within liposomes achieved in the laboratory by dehydration–hydration cycles mimicking those in high intertidal pools.

27. Naylor, E. (2010). *Chronobiology of marine organisms*. Cambridge University Press. Review of the history of tides and of plant and animal adaptations to tides by acquisition of inherited biological clocks.

28. Jones, S. (1999). *Almost like a whale*. Doubleday. A modern view of the origin of species with reference to the higher than present estimated speed of the Earth's rotation in the Palaeozoic era.

29. Palmer, J. D. (1995). *The biological rhythms and clocks of intertidal animals*. Oxford University Press, 5. A pioneering book on the rhythmic adaptations of intertidal animals to the rise and fall of tides.

30. Zhang, L., Hastings. M. H., Green, E. W., Tauber, E., Sladek, M., Webster, S. G., Kyriacou, C. P. and Wilcockson, D. C. (2013). Dissociation of circadian and circatidal timekeeping in the marine crustacean *Eurydice pulchra*. *Current Biology*, 23, 1–11. First molecular evidence that circatidal biological clocks are distinct from circadian clocks.

31. Zantke, J., Ishikawa-Fujiwara, T., Arboleda, E., Lohs, C., Schipany, K., Hallay, N., Straw, A. D., Todo, T. and Tessmar-Raible, K. (2013). Circadian and circalunar clock interactions in a marine annelid. *Cell Reports*, 5, 99–113. First molecular evidence that distinct circalunar biological clocks are recognizable within an organism.

32. Williams, J. L. (1898). Reproduction in *Dictyota dichotoma*. *Annals of Botany*, 12, 559–60. This and the next entry are two early and convincing reports of lunar periodicity of reproduction in a marine alga.

33. Williams, J. L. (1905). Studies on the Dictyotaceae III. The periodicity of the sexual cells of *Dictyota dichotoma*. *Annals of Botany*, 19, 531–60.

34. Muller, D. (1962). Uber jahres und lunarperiodische erscheinungen bei einigen braunalgen. *Botanica Marina*, 4, 140–55. Confirmation by laboratory experiments that a seaweed exhibits Moon-related reproduction.

35. Foster, R. and Kreitzman, L. (2004). *Rhythms of life*. Profile Books. An authoritative book by eminent chronobiologists, demonstrating that circadian rhythms and clocks control the daily lives of 'every living thing', with implications for human ways of life and for human medicine.

36. Brown, F. A., Jr. (1954). Persistent activity rhythms in the oyster. *American Journal of Physiology*, 178, 510–14. Oysters transported inland respond to lunar gravity. Or do they?

37. Enright, J. T. (1965). The search for rhythmicity in biological time-series. *Journal of Theoretical Biology*, 8, 426–68. A critical search for objective methods to analyse for rhythmicity in time-series of biological data.

38. Cole, L. C. (1957). Biological clock in the unicorn. *Science*, 125, 874. An unsubtle challenge to some early reports of biological rhythms, which were purported to correlate with lunar variables.

39. Gamble, F. W. and Keeble, F. (1903). The bionomics of *Convoluta roscoffensis*, with special reference to its green cells. *Proceedings of the Royal Society of London, B*, 72, 93–8. This and the following three papers provide probably the first evidence of tidal biological clocks in a marine animal.

40. Bohn, G. (1903). Sur les mouvements oscillatoires des *Convoluta roscoffensis*. *Comptes rendue Académie des Sciences, Paris*, 137, 576–8.

41. Bohn, G. and Pieron, H. (1906). Le rhythme des marées et la phénomène de l'anticipation réflexe. *Comptes rendues Société de Biologie, Paris*, 61, 660–1.

42. Martin, L. (1907). La mémoire chez *Convoluta roscoffensis*. *Compte rendue hebdomadaire de séances de l'Académie des Sciences*, 145, 555–7.

43. Brown, F. A., Jr. (1962). *Biological clock*. American Institute for Biological Sciences, B.S.C.S. Pamphlet 2, Heath and Co., Boston,

MA. Presents an early argument that tidal rhythms of behaviour in animals are partly controlled by small-scale external environmental variables, not by internal clocks.

44. Naylor, E. (2001). Marine animal behaviour in relation to lunar phase. *Earth, Moon and Planets*, 85–6, 291–302. An academic presentation of evidence that some marine animals exhibit behaviour patterns which vary according to particular phases of the Moon.

45. Klapow, L. A. (1976). Lunar and tidal rhythms of an intertidal crustacean. In: P. J. DeCoursey (ed.), *Biological rhythms in the marine environment*, University of South Carolina Press, Columbia, SC, 215–24. An impressive example of clock-controlled tidally related, up- and down-shore migrations in an intertidal crustacean from the west coast of the USA.

46. Reid, D. G. and Naylor, E. (1986). An entrainment model for semilunar rhythmic swimming behaviour in the marine isopod *Eurydice pulchra. Journal of Experimental Marine Biology and Ecology*, 100, 25–35. Seeks to answer the question as to whether the biological clock of fortnightly (semilunar) periodicity in a sand-beach crustacean is synchronized directly or indirectly by the Moon.

47. Munro Fox, H. (1928). *Selene, or sex and the moon*. Kegan and Paul, Trench, Trubner and Co. An early short volume concerned with the application of science to fable and folklore, seeking to enhance understanding of the Moon's influence on living organisms.

48. Morgan, E. (2001). The moon and life on earth. *Earth, Moon and Planets*, 85–6, 279–90. The article states: 'The robust regulation of animal behaviour by the ocean tides contrasts sharply with the evidence for periodic lunar intervention in human behaviour and physiology.' It concludes: 'the argument for further empirical study of the Moon's influence on life on Earth is substantial.'

49. Lamare, M. D. and Stewart, B. G. (1998). Full moon spawning of *Evechinus* in Doubtful Sound. *Marine Biology*, 132, 135–40. This and the following two papers describe Moon-related spawning in sea urchins in New Zealand, the USA, and Japan.

50. Kennedy, B. and Pearse, J. S. (1975). Lunar synchronization of the monthly reproductive rhythm in the sea urchin *Centrostephanus*

coronatus Verrill. *Journal of Experimental Marine Biology and Ecology*, 17, 323–31.

51. Kobayashi, N. (1967). Spawning periodicity of sea urchins at Seto, I. *Mespilia globalus*. *Publications of the Seto Marine Biological Laboratory*, 14/5, 403–14.

52. Korringa, P. (1957). Lunar periodicity. *Memoirs of the Geological Society of America*, 67, 917–34. A comprehensive review which documents many examples of lunar periodicity, expressing scepticism that some such rhythms might be controlled directly by variations in moonlight intensity.

53. Grahame, J. and Branch, G. M. (1985). Reproductive patterns of marine invertebrates. *Oceanography and Marine Biology Annual Reviews*, 23, 373–98. Article demonstrating that marine snails release larvae in a semilunar pattern, even when kept in the laboratory.

54. Caspers, H. (1984). Spawning periodicity and habitat of the palolo worm (*Eunice viridis*) in the Samoan Islands. *Marine Biology*, 79, 229–36. Documentation of a historical record of Moon-related spawning dates of Samoan palolo from 1843 until the early 1980s.

55. Bentley, M. G., Olive, P. J. W. and Last, K. (2001). Sexual satellites, moonlight and nuptial dances of worms: the influence of the moon on the reproduction of marine animals. *Earth, Moon and Planets*, 85–6, 67–84. A comprehensive study of the influence of the Moon on the reproduction of marine animals, particularly marine worms.

56. Hauenschild, C. (1960). Lunar periodicity. *Cold Spring Harbor Symposia on Quantitative Biology*, 25, 491–7. A pioneering experimental study of lunar periodicity of reproduction in the bristle worm *Platynereis dumerilii*.

57. Winfree, A. T. (1987). *The timing of biological clocks*. Scientific American Library, 19, Scientific American Books, Inc., New York, 7. An engineer looks at chronobiology and finds beauty in the mathematics of biological timing.

58. Hayes, G . C., Akesson, S. and Broderick, A. C. (2003). Island-finding ability of marine turtles. *Proceedings of the Royal Society of London, B*, 270, Supp.1, 5. How turtles find their way to their breeding beaches on remote islands.

59. Barnes, R. D. (1987). *Invertebrate Zoology*, Saunders College Publishing, Philadelphia. The structure and life history of horseshoe crabs described in detail in a classical textbook of invertebrate zoology.

60. Enright, J. T. (1975a). Orientation in time: endogenous clocks. In *Marine Ecology*, 2/2, Physiological Mechanisms, ed. O. Kinne. Wiley Interscience, 914–44. A short review of grunion spawning, based partly on a doctoral thesis by B. W. Walker (1949) at the University of California, Los Angeles.

61. Hartnoll, R. G., Broderick, A. C., Godley, B. J., Musick, S., Pearson, M., Stroud, S. A. and Saunders, K. E. (2010). Reproduction in the land crab *Johngarthia lagostoma* on Ascension Island. *Journal of Crustacean Biology*, 30/1, 83–92. An article that describes Moon-related periodicity of times of marching by land crabs on their way to spawn at the edge of the sea.

62. Hicks, J. W. (1985). The breeding behaviour and migration of the terrestrial crab *Geocarcoidea natalis* (Decapoda: Brachyura). *Australian Journal of Zoology*, 33, 127–42. Another example of a land crab that marches to the edge of the sea to spawn, phased by the Moon.

63. Saigusa, M. (1982). Larval release rhythm coinciding with solar day and tidal cycles in the terrestrial crab *Sesarma*: harmony with the semilunar timing and its adaptive significance. *Biological Bulletin*, 162, 371–86. An account of Moon-related spawning in Japanese land crabs.

64. Saigusa, M. (1988). Entrainment of tidal and semilunar rhythms by artificial moonlight cycles. *Biological Bulletin*, 174/2, 126–38. Experimental demonstration that exposure to cycles of artificial moonlight re-sets the times of spawning by land crabs.

65. Neumann, D. (1965). Photoperiodische steuerung de 15-tagigen lunaren metamorphose periodic von *Clunio* population (Diptera: Chironomidae). *Zeitschrift für Naturforschung*, 206, 818–19. Presents clear evidence that exposure to moonlight induces accurate swarming times of a marine midge.

66. Neumann, D. (1987). Tidal and lunar adaptations of reproductive activities in invertebrate species. In: L. Pevet (ed.), *Comparative physiology of environmental adaptations* III (Karger, Basel), 152–70.

A comprehensive review of tidal and Moon-related breeding patterns in a number of invertebrate animals.

67. Phillips, B. F. (1981). The circulation of the south-eastern Indian Ocean and the planktonic life of the western rock lobster. *Oceanography and Marine Biology: Annual Reviews*, 19, 11–39. An account of the dispersal of rock lobster larvae into the Indian Ocean and their return to lobster fishing grounds in western Australia.

68. Skov, M. W., Hartnoll, R. G., Ruwa, R. K., Shunula, J. P., Vannini, M. and Cannicci, S. (2005). Marching to a different drummer: crabs synchronize reproduction to a 14-month lunar-tidal cycle. *Ecology*, 86/5, 1164–71. A remarkable example of adaptation by crabs to the 14-month lunar cycle during which highest tides alternate between new and full Moon spring tides.

69. Zeng, C., Abello, P. and Naylor, E. (1999). Endogenous tidal and semilunar moulting rhythms in early juvenile shore crabs, *Carcinus maenas*: implications for adaptation to a high intertidal habitat. *Marine Ecology Progress Series*, 191, 257–66. Explains how crab larvae in the plankton display Moon-related behaviour, permitting them to settle out and moult in their optimal zone for development high on the shore.

70. Williams, J. A. (1979). A semilunar rhythm of locomotor activity and moulting synchrony in the sand-beach amphipod *Talitrus saltator*. *European Marine Biology Symposia*, 13, 407–14. Explains how behaviour indirectly related to the Moon ensures that newly hatched sandhoppers avoid desiccation by emerging on to the beach surface when highest levels of the beach are reached by the tide.

71. Metcalfe, J. D., Hunter, E. and Buckley, A. A. (2006). The migratory behaviour of North Sea plaice: currents, clocks and cues. *Marine and Freshwater Behaviour and Physiology*, 39/1, 25–36. Describes techniques for tracking juvenile plaice as they exploit alternating tidal currents when they return to the North Sea locality where they were spawned.

72. Hsiao, S. M. and Meier, A. H. (1989). Comparison of semilunar cycles of spawning activity in *Fundulus grandis* and *F. heteroclitus* held under constant laboratory conditions. *Journal of Experimental*

Zoology, 252, 213–18. Laboratory studies demonstrating Moon-related cycles of spawning in fish.

73. Linkowski, T. B. (1996). Lunar rhythms of vertical migrations coded in otolith microstructures of North Atlantic lantern fishes, genus *Hygophum* (Myctophidae). *Marine Biology*, 124, 495–508. Evidence of natural data-logging by lantern fish as they undertake Moon-related rhythms of upwards swimming in the sea.

74. Farbridge, K. J. and Leatherland, J. F. (1987). Lunar periodicity of growth cycles in rainbow trout *Salmo gairdneri* Richardson. *Journal of Interdisciplinary Cycle Research*, 18, 169–77. Reports how the timing of growth cycles of trout varies with the Moon.

75. Corbet, P. S. (1958). Lunar periodicity in aquatic insects in Lake Victoria. *Nature*, 182, 330–1. Describes how insects in an African lake show varied behaviour in relation to the Moon.

76. Rossiter, A. (1991). Lunar spawning synchroneity in a freshwater fish. *Naturwissenschaften*, 78, 182–4. The first comprehensive account of Moon-related spawning in lake-dwelling fish.

77. Sponaugle, S. and Pinkard, D. (2004). Lunar cyclic population replenishment of a coral reef fish: shifting patterns following oceanic events. *Marine Ecology Progress Series*, 267, 267–80. A well-argued case suggesting how lunar periodicity of behaviour in a marine fish may have evolved in relation to the periodicity of tides.

78. Herring, P. J. and Roe, H. S. J. (1988). The photoecology of pelagic oceanic decapods. *Symposia of the Zoological Society of London*, 59, 263–90. A comprehensive review describing how the day/night cycle and the lunar monthly cycle of light intensity affect the vertical migration behaviour of some ocean plankton communities.

79. Denton, E. J. (1970). On the organization of reflecting surfaces in some marine animals. *Philosophical Transactions of the Royal Society, B*, 225, 63–97. Study showing how reflecting surfaces of fish and other marine animals help to confuse predators.

80. Aldredge, A. L. and King, J. M. (1980). Effects of moonlight on vertical migration patterns of demersal plankton. *Journal of Experimental Marine Biology and Ecology*, 44/2, 133–56. Explains how

moonlight affects the behaviour of animals that migrate between the sea-bed and the water column above.

81. Trillmich, F. and Mohren, W. (1981). Effects of the lunar cycle on the Galapagos fur seal, *Arctocephalus galapagoensis. Oecologia (Berlin)*, 48, 85–92. Do fur seals in the Galapagos Islands come ashore during full Moon to avoid predatory sharks or because of scarcity of food?

82. Cheng, L. and Enright, J. T. (1973). Can *Halobates* dodge nets? II. By moonlight? *Limnology and Oceanography*, 18, 666–9. A study by oceanographers asking whether ocean-skating insects are able to avoid capture in plankton nets during nights of the full Moon.

83. Mitchell, B. and Hazlett, B. A. (1996). Predator avoidance strategies of the crayfish *Orconectes virilis. Crustaceana*, 69/3, 400–12. Crayfish in rivers, in the absence of tides, are shown to avoid predators during bright moonlight.

84. Kotler, B. P., Brown, J., Mukherjee, S., Berger-Tal, O. and Bouskila, A. (2010). Moonlight avoidance in gerbils reveals a sophisticated interplay among time-allocation, vigilance and state-dependent foraging. *Proceedings of the Royal Society of London, B*, 277 (1687), 1469–74. Avoidance of moonlight by gerbils depends on a complicated interplay of time-allocation, vigilance, and hunger state.

85. O'Connell, S. (1992). Desert rats prefer twilight zones. *New Scientist*, 135, 17. Desert rats are observed to forage above ground at night, but not during full Moon.

86. Goodenough, J. E., McGuire, B. and Wallace, R. A. (1993). *Perspectives in Animal Behaviour*, John Wiley and Sons. Describes remarkable Moon-related changes in the size of the sand-trap built by the ant-lion, a predatory insect.

87. Griffin, P. C., Griffin, S. C., Waroquiers, C. and Mills, L. C. (2005). Mortality by moonlight: predation risk and the snowshoe hare. *Behavioural Ecology*, 16, 938–44. Snowshoe hares are secretive at night but only on moonlit snowy nights.

88. Mougeot, F. and Bretagnolla, V. (2000). Predation risk and moonlight avoidance in nocturnal seabirds. *Journal of Avian Biology*, 31, 376–86. Some nocturnal seabirds are shown to reduce the risk of

predation by other nocturnal birds by avoiding flight during bright moonlit nights.

89. Packer, C., Ikanda, D., Kisrui, B. and Kushnir, H. (2005). Lion attacks on humans in Tanzania. *Nature*, 436, 927–8. A scientific article which generated a newspaper headline: 'Beware a full Moon, it's a sign you may soon be a lion's lunch!'

90. Chapin, J. P. and Wing, L. W. (1959). The Wideawake Calendar 1941–1959. *Auk*, 76, 153–8. Describes an early version of the 'Wideawake Calendar' showing Moon-related timing of breeding by a marine bird.

91. Enright, J. T. (1975b). The moon illusion examined from a new point of view. *Proceedings of the American Philosophical Society*, 119/2, 87–107. A useful introduction to the Moon Illusion.

92. Kaufman, L. and Kaufman, J. H. (2000). Explaining the moon illusion. *Proceedings of the American Philosophical Society*, 97/1, 500–5. Synthesis of explanations for the Moon Illusion, an enigmatic phenomenon that has still to be fully explained.

93. Hoffman, K. (1982). Time-compensated celestial orientation. In: J. Brady (ed.), *Biological timekeeping*, Society for Experimental Biology Seminar Series, 4, 49–62. An authoritative review of orientation and navigation behaviour by animals guided by celestial cues combined with use of their own internal biological clocks.

94. Leschiutta, S. and Tavella, P. (2001). Reckoning time, longitude and the history of the earth's rotation, using the moon. *Earth, Moon and Planets*, 85–6, 225–36. A review of the methods used by early human explorers in the quest to solve the problem of determining longitude, using the Moon.

95. Jager, Z. (1999). *Floundering: process of tidal transport and accumulation of larval flounder* (Platichthys flesus) *in the Ems-Dollard nursery*. Ponsen and Looijen, Wageningen, 192pp. Explains how flounder larvae utilize Moon-driven tides in their journey from the open sea to accumulate on nursery grounds in estuaries along the coastline of the Netherlands.

96. Wood. L. and Hargis, J. H. (1971). Transport of bivalve larvae in a tidal estuary. *European Marine Biology Symposia*, 4, 29–44. A major study explaining how oyster bed integrity is maintained in an

estuary: free-swimming larvae utilize Moon-driven tides to avoid being washed seawards.

97. Epifanio, C. F., Valenti, C. C., and Pembroke, A. E. (1984). Dispersal and recruitment of blue crab larvae in Delaware Bay, USA. *Estuarine, Coastal and Shelf Science*, 18, 1–12. Explains how blue crab populations in a US estuary are recruited back to sustain the parent stock after larvae are dispersed in the open sea.

98. Tankersley, R. A., McKelvey, L. M. and Forward, R. B. (1995). Responses of estuarine crab megalopae to pressure, salinity and light: implications for flood-tide transport. *Marine Biology*, 122, 391–400. An account of the detailed behavioural responses of blue crab larvae which enable them to exploit Moon-driven tidal action as they move from the sea into estuaries.

99. Dacke, M., Nilsson. D. E., Scholtz, C. H., Byrne, M. J. and Warrant, E. I. (2003). Insect orientation to polarized light. *Nature*, 424, 33. Claimed to be the first report of insect navigation by the Moon.

100. Dacke, M., Byrne, M. J., Scholtz, C. H. and Warrant, E. J. (2004). Lunar orientation in a beetle. *Proceedings of the Royal Society, B*, 271, 361–5. More details concerning the discovery of insect navigation by the Moon.

101. Papi, F. and Pardi, L. (1963). On the lunar orientation of sandhoppers (Amphipoda: Talitridae). *Biological Bulletin*, 124, 97–105. An early reference to the possibility that European sandhoppers may be able to orientate their homing migrations in relation to the position of the Moon.

102. Enright, J. T. (1972). When the beach-hopper looks at the moon: the moon-compass hypothesis. In: S. R. Galler et al. (eds), *Animal orientation and navigation* (National Aeronautics and Space Administration, Washington, DC), 523–55. A further early reference, from studies in California, to the possibility that sandhoppers are able to orientate their homing behaviour in relation to the position of the Moon.

103. Ugolini, A., Boddi, V., Mercatelli, L. and Castellani, C. (2005). Moon orientation in adult and young sandhoppers under artificial light. *Proceedings of the Royal Society, B*, 272, 2189–94. First demonstration

by laboratory experiments that sandhoppers are able to navigate by the Moon.

104. Foster, R. G. and Roenneberg, T. (2008). Human responses to the geophysical daily, annual and lunar cycles. *Current Biology*, 18/17, R784–R794. Concludes unequivocally that there is no solid evidence that the human body is regulated by the lunar cycle.

105. Law, S. P. (1986). The regulation of the menstrual cycle and its relationship to the moon. *Acta Obstetrica Gynecologie Scandinavia*, 65, 45–8. A claim that there is a synchronous relationship between the timing of menstruation and the phase of the Moon.

106. Cutler, B. (1980). Lunar and menstrual phase-locking. *American Journal of Obstetrics and Gynecology*, 137, 834–9. Demonstrates a correlation between menstruation and the Moon, in a selected sample of women.

107. Folin, M. and Rizzotti, M. (2001). Lunation and the primate menses. *Earth, Moon and Planets*, 85–6, 539–43. Presents a detailed comparison of the lengths of human and other primate menstrual cycles.

108. Ghiandoni, G., Secli, R., Rocchi, M. B. and Ugolini, G. (1998). Does lunar position influence the time of delivery? A statistical analysis. *European Journal of Obstetrics, Gynecology and Reproductive Biology*, 77, 47–50. An attempt to answer the question as to whether times of childbirth are influenced by the position of the Moon.

109. Periti, E. and Biagiotti, R. (1994). Lunar phases and incidence of spontaneous deliveries: our experience. *Minerva Ginecol*, 46, 429–33. Survey evidence concluding that the Moon does not influence the time of human childbirth.

110. Strestik, J., Sitar, J., Predeanu, I. and Botezat-Antonescu, L. (2001). Variations in mortality with respect to lunar phase. *Earth, Moon and Planets*, 85–6, 567–72. An extensive survey of times of human deaths in relation to lunar phase.

111. Strestik, J. and Sitar, J. (1996). The influence of heliophysical and meteorological factors on cardiovascular mortality. In: *Proceedings of 14th International Congress of Biometeorology, Lubliana*, 166–73. A Czech Republic study of times of death in relation to lunar phase.

112. Ali, Y., Rahme, R., Matar, N., Ibrahim, I., Menassa-Moussa, L., Maarawi, J. et al. (2008). Impact of the lunar cycle on the incidence of intracranial aneurysm rupture: myth or reality? *Clinical Neurology and Neurosurgery*, 110, 462–5. Reports a correlation between subarachnoid aneurysm rupture and lunar phase.

113. Lahner, D., Merhold, F., Gruber, A. and Schramm, W. (2009). Impact of the lunar cycle on the incidence of aneurysmal subarachnoid haemorrhage: myth or reality? *Clinical Neurology and Neurosurgery*, 111, 352–3. Quickly contradicts Ali et al. (2008)—considers the correlation to be a myth!

114. Schuhaiber, J. H., Fava, J., Shin, T., Dobrilovic, N., Ehsan, A., Bert, A. and Selike, R. (2013). The influence of the seasons and lunar cycle on hospital outcomes following ascending aortic dissection repair. *Interactive Cardio-Vascular Thoracic Surgery*, Online, 9 July 2013. Presents evidence to suggest that survival rates after major heart surgery increase if surgery is carried out around the time of full Moon.

115. Cajochen, C., Altanay-Ekici, S., Munch, M., Frey, S., Knoblauch, V. and Wirz-Justice, A. (2013). Evidence that the lunar cycle influences human sleep. *Current Biology*, 23, 1485–8. The first evidence from rigorously controlled experiments that the Moon influences human sleep.

116. Smith, M., Croy, I. and Waye, K. P. (2014). Human sleep and cortical reactivity are influenced by lunar phase. *Current Biology*, 24/12, R551–R552. Confirmatory evidence that the Moon influences human sleep patterns.

117. Cordi, M., Ackermann, S., Bes, F. W., Hartmann, F., Konrad, B. N., Genzel, L., Pawlowski, M., Steiger, A., Schulz, H., Rasch, B. and Dresler, M. (2014). Lunar cycle effects on sleep and the file drawer problem. *Current Biology*, 24/12, R549–R550. A challenge to the view that the Moon influences human sleep patterns.

118. Uglow, J. (2002). *The lunar men*. Faber. A history of the Lunar Society of Birmingham.

119. Thompson, D. A. and Adams, S. L. (1996). The full moon and ED patient volumes: unearthing a myth. *American Journal of Emergency*

Medicine, 14, 161–4. A study seeking to dispel preconceptions that hospital emergency department admissions increase during full Moon.

120. Coates, W., Jehle, D. and Cottington, E. (1989). Trauma and the full moon: a waning theory. *American Emergency Medicine*, 18, 763–5. Dispelling the view that violent human behaviour, evidenced by trauma admissions, correlates with lunar phase.

121. Thakur, C. P. and Sharma, D. (1984). The full moon and crime. *British Medical Journal (Clinical Research Edition)*, 289, 1789–91. Presents some evidence of Moon-related occurrence of crime.

122. Maldonado, G. and Kraus, J. F. (1991). Variation in suicide occurrence by time of day, day of the week, month and lunar phase. *Suicide and Life-Threatening Behavior*, 1, 174–87. A convincing presentation of evidence claiming that suicide rates are not Moon-related.

123. Deantonio, M. (2001). 'Lunacy' in mentally disturbed children. *Earth, Moon and Planets*, 85–6, 129–31. Explains how unfounded suggestions of Moon-related behaviour are used as 'coping mechanisms' to account for symptoms of mental illness, particularly in children.

124. Zurcher, E., Cantiani, M.-G., Sorbetti Guerri, F. and Michel, D. (1998). Tree stem diameters fluctuate with tide. *Nature*, 392, 665–6. A seemingly authoritative claim, subsequently retracted (see no. 16), that the Moon induces 'tides' in the water transport system of trees to the extent that tree stem diameters fluctuate accordingly.

125. O'Neill, J. S., Lee, K. D., Zhang, L., Feeney, K., Webster, S. G., Blades, M. J., Kyriacou, C. P., Hastings, M. H. and Wilcockson, D. C. (2015). Metabolic molecular markers of the tidal clock in the marine crustacean *Eurydice pulchra*. *Current Biology*, 25, R1–R3 (20 April). The first paper to begin to characterize the molecular circatidal clock of a marine crustacean.

126. Palmer, J. D., Udry, J. R. and Morris, N. M. (1982). Diurnal and weekly, but no lunar rhythms in human copulation. *Human Biology*, 54, 111–21. A rare scientific test of a purported relationship between human libido and the moon.

127. Mercier, A., Sun, Z., Baillon, S. and Hamel, J. F. (2011). Lunar rhythms in the deep sea: evidence from the reproductive periodicity of several marine invertebrates. *J. Biol. Rhythms*, 26, 82–6. Reports the discovery of moon-related breeding cycles in some deep sea animals, with speculation as to how they might be controlled.

128. Meschini, E., Gagliardo, A. and Papi, F. (2008). Lunar orientation in sandhoppers is affected by shifting both the moon phase and the daily clock. *Animal Behaviour*, 76, 25–35. Emerging evidence that moon-related biological clocks are used as a basis for lunar navigation by sandhoppers.

INDEX

ARE DOLPHINS REALLY SMART?

The mammal behind the myth

Justin Gregg

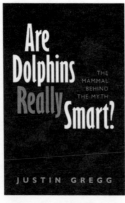

978-0-19-968156-3 | Paperback | £9.99

'Serves as both a rigorous litmus test of animal intelligence and a check on human exceptionalism.' Bob Grant, *The Scientist*

'[T]horough and engaging [Gregg's] writing skills are solid and his observations are often fascinating.' *Booklist*

The Western world has had an enduring love affair with dolphins since the early 1960s, with fanciful claims of their 'healing powers' and 'super intelligence'. Myths and pseudoscience abound on the subject. Justin Gregg weighs up the claims made about dolphin intelligence and separates scientific fact from fiction. He puts our knowledge about dolphin behaviour and intelligence into perspective, with comparisons to scientific studies of other animals, especially the crow family and great apes.

GREEN EQUILIBRIUM

The vital balance of humans and nature

Christopher Wills

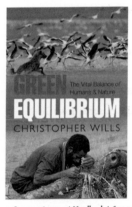

978-0-19-964570-1 | Hardback | £20.00

'*Green Equilibrium* is a richly enlightening exploration of the world and our place in it, infused with fresh insight from science and a deep concern about the future of our planet and our species.'

Carl Zimmer, author of *A Planet of Viruses* and *The Tangled Bank: An Introduction to Evolution*

Across the planet, unique ecosystems are under threat. Overexploitation, pollution, and sheer human population growth put pressure on these delicately balanced 'green equilibria'. Through his vividly described travels, Christopher Wills illustrates the principles of ecology and evolution that underlie the rich mosaics of environments as varied as Californian grasslands, Philippine coral reefs, and the remote mountainous jungle-clad valleys of Papua New Guinea. Using the latest genetic evidence of our evolutionary past, Wills shows how humans form an integral part of the story, and how we are shaped by the ecosystems in which we settled as we spread across the planet. Using a number of striking examples, he demonstrates how we can halt the damage already done, and help preserve the green equilibria for the local communities who have lived and adapted to them.

HOMO MYSTERIOUS

Evolutionary puzzles of human nature

David P. Barash

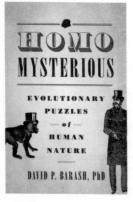

978-0-19-932452-1 | Paperback | £12.99

'[Barash] succeeds admirably in explaining both the pieces of the puzzles and the strengths and weaknesses of solutions to date.'

The Sunday Telegraph

For all that science knows about the living world there are even more things that we don't know, genuine evolutionary mysteries that perplex the best minds in biology. Paradoxically, many of these mysteries are very close to home, involving some of the most personal aspects of being human.

Readers are plunged into an ocean of unknowns—the blank spots on the human evolutionary map, the terra incognita of our own species—and are introduced to the major hypotheses that currently occupy scientists who are attempting to unravel each puzzle (including some solutions proposed here for the first time). Throughout the book, readers are invited to share the thrill of science at its cutting edge, a place where we know what we don't know, and, moreover, where we know enough to come up with some compelling and seductive explanations.

ISLANDS BEYOND THE HORIZON

The life of twenty of the world's most remote places

Roger Lovegrove

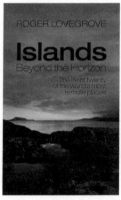

978-0-19-872757-6 | Paperback | £9.99

'Lovegrove manages to capture each island's identity and mystery and transmits his affection for these faraway places.' *Northern Echo*

Islands have an irresistible attraction and an enduring appeal. Naturalist Roger Lovegrove has visited many of the most remote islands in the world, and in this book he takes the reader to the twenty that fascinate him most. Some are familiar but most are little known; they range from the storm-bound island of South Georgia and the ice-locked Arctic island of Wrangel to the wind-swept, wave-lashed Mykines and St Kilda.

The range is diverse and spectacular; and whether distant, offshore, inhabited, uninhabited, tropical or polar, each is a unique self-contained habitat with a delicately-balanced ecosystem, and each has its own mystique and ineffable magnetism. By looking not only at the biodiversity of each island, but also the uneasy relationship between its wildlife and the involvement of man, Lovegrove provides a richly detailed account of each island, its diverse wildlife, its human history, and the efforts of conservationists to retain these irreplaceable sites.

MARINE BIOLOGY

A Very Short Introduction

Philip V. Mladenov

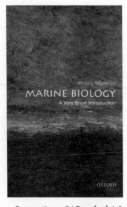

978-0-19-969505-8 | Paperback | £7.99

'The easy accessibility of this book means that readers will take heed. Recommended.'

J. A. Mather, *Choice*

The marine environment is the largest, most important, and yet most mysterious habitat on our planet. It contains more than 99% of the world's living space; produces half of its oxygen; plays a critical role in regulating its climate; and supports a remarkably diverse and exquisitely adapted array of life forms, from microscopic viruses, bacteria, and plankton, to the largest existing animals.

In this unique *Very Short Introduction*, Philip Mladenov provides a comprehensive overview of marine biology, providing a tour of marine life and marine processes that ranges from the polar oceans to tropical coral reefs; and from the intertidal to the hydrothermal vents of the deep sea.

Sign up to our quarterly e-newsletter **http://academic-preferences.oup.com/**

THE AMOEBA IN THE ROOM

Lives of the microbes

Nicholas P. Money

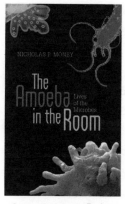

978-0-19-966593-8 | Hardback | £16.99

'An impassioned fascination for microscopic life around and within us... Overall, Money delivers a heady mixture of history, philosophy, art and even poetry... This is a lucid and informative book. There is an impressive afterword of references and notes, and fine line drawings. So much that is lyrical and little-known waits to be discovered here—novelties that will appeal to new undergraduates as well as to incorrigible microbial enthusiasts like myself.'

Mark O. Martin, *Nature*

The more we learn about microbial biodiversity, the less important do animals and plants become in our understanding life on earth. The flowering of microbial science is revolutionizing biology and medicine in ways unimagined even a decade or two ago, and is inspiring a new view of what it means to be human. Nicholas P. Money explores the extraordinary breadth of the microbial world and the vast swathes of biological diversity that are now becoming recognized using molecular methods.

THE DANCE OF AIR AND SEA

How oceans, weather, and life link together

Arnold H. Taylor

978-0-19-956559-7 | Hardback | £16.99

'This is a fascinating tale of a continuing journey toward understanding a complex system, and of the changes that the thickening greenhouse gas blanket may cause to it.'

> Robert May, former President
> of The Royal Society

How can the tiny plankton in the sea just off Western Europe be affected by changes 6000 km away on the other side of the North Atlantic Ocean? How can a slight rise in the temperature of the surface of the Pacific Ocean have a devastating impact on amphibian life in Costa Rica?

In *The Dance of Air and Sea* Arnold H. Taylor focuses on the large-scale dynamics of the world's climate, looking at how the atmosphere and oceans interact, and the ways in which ecosystems in water and on land respond to changes in weather. He tells stories of how discoveries were made and the scientists who made them, and considers how these crucial issues contribute towards our response to climate change.

Sign up to our quarterly e-newsletter **http://academic-preferences.oup.com/**